THE SHAPE OF A LIFE

SHING-TUNG YAU
AND STEVE NADIS

The Shape of a Life

ONE MATHEMATICIAN'S SEARCH

FOR THE UNIVERSE'S

HIDDEN GEOMETRY

Yale UNIVERSITY PRESS   NEW HAVEN AND LONDON

Published with assistance from the foundation established in memory of William McKean Brown.

Copyright © 2019 by Shing-Tung Yau and Steve Nadis. All rights reserved. This book may not be reproduced, in whole or in part, including illustrations, in any form (beyond that copying permitted by Sections 107 and 108 of the U.S. Copyright Law and except by reviewers for the public press), without written permission from the publishers.

Yale University Press books may be purchased in quantity for educational, business, or promotional use. For information, please e-mail sales.press@yale.edu (U.S. office) or sales@yaleup.co.uk (U.K. office).

Illustrations on pages 49, 58, 78, 81, 101, 119, 176, 178, 199, 205, 234, and 235 are courtesy of Barbara Schoeberl, Animated Earth, LLC.

Set in Scala and Scala Sans type by Integrated Publishing Solutions. Printed in the United States of America.

Library of Congress Control Number: 2018953465
ISBN 978-0-300-23590-6 (hardcover : alk. paper)

A catalogue record for this book is available from the British Library.

This paper meets the requirements of ANSI/NISO Z39.48-1992 (Permanence of Paper).

10 9 8 7 6 5 4 3 2 1

To Our Parents:
Yeuk Lam Leung and Chen Ying Chiu
Lorraine B. Nadis and Martin Nadis

ON THE CENTENNIAL BIRTHDAY OF MY LATE FATHER

An inspiring life of ups and down, vanquished in a moment.
Though his wisdom of East and West still echoes in my heart.
I never enjoyed his love enough, which has left me in dismay.
The bloom of youth has passed me by, my hair turned to gray.
Yet I oft look back to that fateful time when I was just a careless teen.
How sad it was when he left that night, so long ago and faraway.
What might he have told us, I wonder, if only he could have said?
Though I'll never hear those words, his thoughts are with me, always.

—*Shing-Tung Yau, 2011*

CONTENTS

Preface ix

1 Itinerant Youth 1

2 Life Goes On 26

3 Coming to America 44

4 In the Foothills of Mount Calabi 75

5 The March to the Summit 96

6 The Road to Jiaoling 117

7 A Special Year 142

8 Strings and Waves in Sunny San Diego 170

9 Harvard Bound 189

10 Getting Centered 211

11 Beyond Poincaré 233

12 Between Two Cultures 261

Epilogue 277

Index 283

Photographs follow p. 141

PREFACE

Having no prior experience in committing "the story of my life" to the printed page, I'll try to keep things simple—for my sake, if not for yours—and start at the beginning. I was born in China in the spring of 1949 in the midst of the Communist revolution. A few months later, my family moved to Hong Kong, where I lived until going to the United States for graduate school in 1969. In the nearly five decades that have elapsed since my first transpacific crossing, I have gone back and forth between America and Asia on countless occasions. At times, it is hard for me say which is my true home or whether it would be more accurate to say that I have two homes, neither of which I'm fully at home in.

To be sure, I have carved out a comfortable existence in America without ever feeling truly at one with the society around me. I also have strong emotional and familial ties with China that are deeply engrained and seemingly hardwired into my being. Nevertheless, after many decades away, my perspective on my native land has shifted as if I were always observing things from at least one or two steps removed. Whether I'm in America or in China, I feel as if I have both an insider's view and an outsider's view at the same time.

This sense has left me occupying a rather peculiar place that cannot be located on a conventional map—a place lying somewhere between two cultures and two countries that are separated from each other historically, geographically, and philosophically—and through rather profound differences in cuisine. I have a home in Cambridge, Massachusetts, not far from Harvard University, which I'm happy to say has been my employer since 1987. I also have an apartment in Beijing, which I'm delighted to make use of when I'm in town. But there is a third home I've had much longer, and that is mathematics—a field I have been fully ensconced in for almost a half century.

For me, mathematics has offered a kind of universal passport that has allowed me to move freely throughout the world at the same time I ply its formidable tools toward the task of making sense of that world. I've always found mathematics to be a fascinating subject with seemingly magical properties: It can bridge gaps of distance, language, and culture, almost instantly bringing onto the same page—and onto the same plane of understanding—people who know how to harness its power. Another thing that's magical about mathematics is that it doesn't take much, if any, money to do something significant in the field. For many problems, all you need is a piece of paper and a pencil, along with the ability to focus the mind. And sometimes you don't even need paper and pencil—you can do the most important work in your head.

I feel lucky that ever since finishing graduate school, and even before obtaining my PhD, I have never stopped pursuing research in my chosen field. Along the way, I've made some contributions to this discipline that I'm proud of. But a career in mathematics was by no means assured for me, despite a fascination with the subject that took hold of me during childhood. In fact, early in my life, the path I currently find myself on appeared to be well beyond reach.

I grew up poor in terms of the standard financial metrics but rich in the love my mother and father bestowed upon my siblings and me, and in the intellectual nourishment we received. Sadly, my father, Chen Ying Chiu, died when I was just fourteen years old, throwing our family into dire economic straits—with no "nest egg" to fall back on and mounting debts from all sides. My mother, Yeuk Lam Leung, was nonetheless determined for us to continue our education—a wish that was consonant with that of my father, who had always encouraged us toward scholarly pursuits. I became serious about my studies and found my calling in mathematics—a subject I was drawn to in middle school and high school in Hong Kong.

A big break came during my college years in Hong Kong upon meeting Stephen Salaff, a young mathematician from the University of California, Berkeley. Salaff arranged for me to pursue graduate studies at Berkeley, enlisting the services of a powerful member of the school's math department, Shiing-Shen Chern, who was then the world's foremost mathematician of Chinese descent.

I don't know whether I would have gotten far in my field had it not been for the fortuitous chain of events that brought me to California. But I am certain of one thing: I never would have been able to secure such a career had it not been for the sacrifices that my mother made for all of her children and for the love of learning that my father instilled in all of his progeny. I dedicate this book to my parents, who made it possible for me to live out the story told here. I also thank my wife Yu-Yun and my sons, Isaac and Michael, who have put up with me over the past several decades, and to all of my brothers and sisters.

I have spent innumerable hours indulging my obsession for shapes and numbers, as well as for curves, surfaces, and spaces of any dimension. But my work, as well as my life, has also been enriched, immeasurably so, by my relationships with people—family, friends, colleagues, professors, and students.

This is the story of my odyssey—between China, Hong Kong, and the United States. I have traveled the world in my pursuit of geometry—a field that is crucial to our attempts to map out the universe on both the largest and smallest scales. Conjectures have been made during these excursions, "open problems" raised, and various theorems proved. But work in mathematics is almost never done in isolation. We build upon history and are shaped by myriad interactions. These interactions can, on occasion, lead to misunderstandings and even fights, which I have, unfortunately, been caught up in from time to time. One of the things I've learned through these incidents is that the notion of "pure mathematics" can be hard to realize in practice. Personalities and politics can intrude in unexpected ways, sometimes obscuring the intrinsic beauty of this discipline.

Nonetheless, chance encounters with peers can also send us in unexpectedly fruitful trajectories that may last years or decades. In the final analysis, we are the products of our times and of our milieus, of whom we come from and where we come from. It now seems as if I come from many places—a fact that has made my life both richer and more complicated. In the account that follows, I hope to convey a sense of my upbringing, growth, and personal journey to any readers who might take an interest.

I take this opportunity to thank some of the many people who—if not contributing to this book directly—helped make the narrative arguably worth telling. For starters, I owe an incalculable debt to my parents, who supported my siblings and me as best they could, through hard times, while always trying to teach us good values. The main purpose of life, I learned, is not about making money—a lesson that enabled me to pursue a career in mathematics rather than in, say, business or banking. I was close to all of my siblings but am especially grateful to my older sister, Shing-Yue, who, up to the moment of her death, sacrificed so much—foregoing a professional career of her own—in order to help me and her other brothers and sisters.

I was also lucky to have fallen in love with, and eventually married, a woman who shared my view that there is more to life than seeking personal wealth, material possessions, and luxuries—that greater rewards can come from scholarly endeavors. I'm proud to see that our sons have also ventured far along academic paths.

I'm lucky to have lifelong friends, like Shiu-Yuen Cheng, Siu-Tat Chiu, and Bun Wong, whom I've known since my school days in Hong Kong. One grade school teacher, Miss Poon, stands out for the kindness she bestowed upon me when I was young and vulnerable. I got an early taste of mathematics from the lecturer H. L. Chow during my freshman year at Chung Chi College. And I was extraordinarily fortunate to have crossed paths during college with Stephen Salaff, who guided me to Berkeley with the help of Chern, Shoshichi Kobayashi, and Donald Sarason.

I'm grateful to the American educational system for providing, since the moment of my arrival, a wonderful environment for pursuing mathematical research. A great feature of this system is that it recognizes and fosters a person's talent, regardless of his or her race, background, or accent. I should single out Harvard in this regard, which has served as a convivial home for me over the past thirty-plus years. I've had many terrific colleagues in the Harvard Mathematics Department—too many in that time, unfortunately, to list here.

My career has been aided immeasurably by somewhat older and more established mathematicians who've gone out of their ways to help me. First and foremost is my former advisor and mentor S. S. Chern. But many others have been of tremendous help, including Armand Borel, Raoul Bott, Eugenio Calabi, Heisuke Hironaka, Friedrich Hirzebruch, Barry Mazur, John Milnor, Charles Morrey, Jürgen Moser, David Mumford, Louis Nirenberg, Robert Osserman, Jim Simons, Isadore Singer, and Shlomo Sternberg.

Some mathematicians prefer to work alone, but I do best in the company of friends and colleagues. I am pleased to have had some great ones over the years, among them S. Y. Cheng, John Coates, Robert Greene, Dick Gross, Richard Hamilton, Bill Helton, Blaine Lawson, Peter Li, Bill Meeks, Duong Phong, Wilfried Schmid, Rick Schoen, Leon Simon, Cliff Taubes, Karen Uhlenbeck, Hung-Hsi Wu, Horng-Tzer Yau, and my brother Stephen Yau. I've collaborated closely, in particular, with Rick Schoen for about forty-five years and have done some of my best work with him. Although he started out as my student, I'm sure I've learned as much from him as he has from me. I truly value his friendship.

I continue to collaborate with other former students and postdocs—such as Huai-Dong Cao, Conan Leung, Jun Li, Bong Lian, Kefeng Liu, Melissa Liu, and Mu-Tao Wang. I've got some outstanding math colleagues in China and Hong Kong: Yang Lo, Zhouping Xin, and many others. I've also had close ties with physicists for most of my career, enjoying my interactions with people like Philip Candelas, Brian Greene, David Gross, Stephen Hawking, Gary Horowitz, Andy Strominger, Henry Tye, Cumrun Vafa, and Edward Witten. My work in mathematics has definitely profited from these associations, and I'd like to think that some benefits have trickled down to physics as well.

All told, it's been an exciting journey so far, and I hope (and expect) there will be a few pleasant surprises on the road ahead.

Shing-Tung Yau
Cambridge, 2018

I have compiled a fair number of publications over the years, including profiles of many individuals, but I've never written a full-length biography before. Frankly, it's been a fascinating experience to plumb the depths of someone's personal history as thoroughly and deeply as one can productively go, and I hope some of that fascination rubs off on those perusing these pages. The task is comparable in some ways to both mining and archaeology—unearthing more and more material, the deeper one digs, and then sifting through the bulk to find the rare gems and other pieces worth holding on to. There are many new things to be learned in the course of such a process, even when the subject of your inquiries is someone you have known, worked closely with, and become friends with for well over a decade.

Of course, I could not have completed this effort without the help of numerous people, and I'd like to thank as many of them as I can, apologizing for any names that I have neglected to mention.

Since this book has a lot to do with family (my coauthor's family, though not mine), I start off by thanking my parents, my wife Melissa Burns—who provided thoughtful feedback on the first three chapters and endured more talk about this book, and its writing, than almost any other human could tolerate—along with my delightful daughters, Juliet and Pauline. I'm also lucky to have two great siblings, my sister Sue and my brother Fred.

My coauthor and I appreciate the unwavering support of our editor, Joe Calamia, and his colleagues at Yale University Press, including Eva Skewes and Ann-Marie Imbornoni. Joe has been encouraging from the very start, maintaining enthusiasm and general cheeriness throughout the long (and sometimes trying) process. Jessie Dolch provided expert copyediting, deftly curbing our tendencies toward (*not* "towards") verbosity, redundancy, and occasional lapses into obscurity. I learned that—regardless of the time, place, or weather—I tend to say "if" when I should say "whether." And I often say "coming" when, to paraphrase Groucho Marx, I should say "going."

The following people also helped with various aspects of the book and my work on it:

Maureen Armstrong	Sergiu Klainerman	Barbara Schoeberl
Lydia Bieri	Joe Kohn	Rick Schoen
Jean-Pierre Bourguignon	Sarah Labauve	Christina Sormani
Maury Bramson	Blaine Lawson	J. Michael Steele
Alicia Burns	Claude LeBrun	Martha Stewart
Huai-Dong Cao	Jun Li	Andy Strominger
Lennart Carleson	Bong Lian	Lydia Suffiad
Lily Chan	Kefeng Liu	Li-Sheng Tseng
Raymond Chan	Yang Lo	Karen Uhlenbeck
S. Y. Cheng	L. Mahadevan	Emmanuel Ullmo
Isaac Chiu	Francisco Martin	Yifang Wang
Siu-Tat Chui	Alex Meadows	Hung-Hsi Wu
Robert Connelly	Bill Meeks	Hao Xu
Daniel Ford	John Milnor	Hongwei Xu
Robert Greene	Irene Minder	Horng-Tzer Yau
Xianfeng (David) Gu	K. F. Ng	Stephen Yau
Simon Guest	Ping-Zen Ong	Xiaotian (Tim) Yin
Richard Hamilton	Dick Palais	Cosmas Zachos
Jennifer Hinneburg	Duong Phong	Chiyuan Zhang
Thomas Hou	Robert Sanders	Lei Zhang
Lizhen Ji	Wilfried Schmid	Xi-Ping Zhu

Maureen Armstrong, who works on the *Journal of Differential Geometry* from within the Harvard Mathematics Department, helped out in many ways—by gathering and preparing the photographs that appear in this book and also by helping to put our manuscript into a presentable form. I am grateful for her efforts and am not sure what we would have done without her. Our deepest gratitude also goes to Lily Chan, who kindly provided many photos along with other assistance. Huai-Dong Cao, Yang Lo, Hao Xu, Hongwei Xu, and Stephen Yau were incredibly helpful. And we heartily thank Xiaotian (Tim) Yin, Xianfeng (David) Gu, and especially Barbara Schoeberl for providing us with some wonderful illustrations. Barbara put all the figures together in only about two weeks—an impressive feat. Andy Hanson also lent some great visualizations of Calabi-Yau manifolds, while offering crucial advice regarding the cover design.

The Berkeley mathematician Hung-Hsi Wu carefully read through each and every chapter draft—going through multiple iterations in some cases. The insights he offered us—about China, the mathematics world, and ways of explaining some complicated mathematical concepts—were invaluable. I am still not sure how he managed to devote so much time to this project, in view of his own appreciable workload, but I'm thankful that he did. And I'm sure that our book is immeasurably better as a result of his sage advice, beneficial prodding, and saintlike patience.

Thank you, Professor Wu, and thanks to everyone else who contributed to this several-year-long undertaking. Sometimes it takes a village, they say. And sometimes it takes even more.

Steve Nadis
Cambridge, 2018

THE SHAPE OF A LIFE

CHAPTER ONE

Itinerant Youth

WE COME ONTO THIS EARTH having no clue as to what life has in store for us—where we'll go, what we'll do, and who we'll become. Some folks, in answer to the first question, live out their days close to where they started, not venturing far from their place of birth. Others cover more ground, and I fall into the latter category, having traveled far and wide within the fields of mathematics and physics, as well as throughout the world at large.

Wanderlust may be my destiny, as well as an engrained part of my heritage, as my family and I are of Hakka extract—an ethnic group thought to have originated in the Yellow River Valley of northern China, moving south during a series of forced migrations over the past one thousand years or so and gradually spreading out from there. Sun Yat-sen, the first president of the Republic of China, and Deng Xiaoping, the most powerful figure in China during the past two decades of the twentieth century, are both of Hakka descent, as was Lee Kuan Yew, the first prime minister and "founding father" of Singapore.

The Hakka people, of whom there are about eighty million today, were originally referred to as "guest people" or "shack people"—wanderers due to necessity rather than a nomadic predisposition. They moved when they had to in order to escape war and famine or, less dramatically, to search for steady employment. The Hakkas endured countless hardships along the way, which became part of their credo, though many clung to the dream of returning someday to their native land. But they also stayed put when the opportunity arose. My ancestors, for example, lived stably in my family's hometown for more than eight hundred years.

However, when Hakka people did settle in one place for a while, they were often consigned to the poorest farmland available, up in the highlands rather than in the fertile valleys below, which had been claimed long before. Up in the dryer, nutrient-deficient soil, farmers were unable to grow China's main crops, rice and wheat, on a successful, large-scale basis and often had to try to cultivate maize and sweet potatoes instead, until even those secondary crops failed. The marginal quality of the land they inhabited might have made the parting easier when the Hakkas were forced to move, once again, because of invasions and other exigencies.

I see some parallels in my own experience. I've been through a number of moves myself, both as a child, when circumstances compelled my family to change venues, and as an adult, where occasional geographic shifts are the norm in academia. I was born in the southern Chinese town of Swatow, now more commonly known as Shantou, on April 4, 1949, the fifth of what would ultimately be eight children. At the time of my birth, I had three older sisters—Shing-Shan, Shing-Hu, and Shing-Yue—and an older brother, Shing-Yuk. My parents carted all five of us off to Hong Kong about six months later, just before the Communists completed their takeover of the government. Hong Kong was a popular haven among intellectuals seeking refuge at the time.

My father, Chen Ying Chiu, shared the widely held view that our Hong Kong sojourn would be a temporary one—that the Communist regime would not last long—a belief history has shown to be erroneous. Some members of my immediate family later ventured to North America or the United Kingdom, but none resumed permanent residence in China.

While I was growing up, my father and mother, Yeuk Lam Leung, mainly spoke Hakkanese, a language that is not so widely heard these days. I was also exposed to Mandarin through conversations with my father's students. Outside of the home, I was forced to speak Cantonese in the Hong Kong schools I attended. My father was strongly influenced by Hakka culture, which put a premium on fostering the intellect (though a greater emphasis, unfortunately, was placed on the education of boys than of girls). It was understood that if you study hard and study well, you could have a future. This strategy paid off for him—intellectually, if not financially—as he became a respected scholar, author, and teacher of philosophy, history, literature, economics, and other subjects.

Because of the important place my father held, and still holds, in my life, I too have been strongly influenced by this same culture. I've tried to pass on some of its basic teachings to my sons, Isaac and Michael, while never losing my penchant for travel—sometimes because it's essential for my work and other times because I like to see the world. I've always felt that it's beneficial to keep exposing yourself to new sights and new ideas, both in the academe and far beyond the confines of the "Ivory Tower."

My father made studying hard a priority for his children, as was the case during his childhood too, although amassing the basic supplies needed to support his scholastic efforts was not easy. He grew up on a farm in Jiaoling County of Guangdong Province, which lies in the southeastern corner of China. His family was so poor they didn't have enough money to buy paper upon which to write. They went to Buddhist temples to collect paper that was typically reserved for religious rituals, putting it to a different use instead—my father's educational pursuits, at which he excelled.

When he was five, he memorized long passages from the *Lunyu*, a collection of teachings from the ancient Chinese philosopher Confucius, also committing to memory tracts from the *Mengzi*, which contained the work of a Confucius follower, the philosopher Mencius. After enrolling in a modern school at the age of seven, my father routinely ranked at the top of his class through high school. When he was eighteen, he went to military school but didn't stay there long, owing to poor health. He later attended Waseda University in Japan, from which he graduated with a master's degree at the age of twenty-two.

My mother was less fortunate in this regard, not having the chance to continue her studies beyond high school, where she worked as a librarian after graduation. (Her father—my grandfather—however, was an esteemed scholar, known for his work in painting, poetry, and calligraphy. He trained several well-known artists, including Lin Fengmian, one of the leading Chinese painters of the twentieth century.) It's worth pointing out that at the time my mother might have gone to college—in the late 1930s—it was rare in China, as well as in other parts of the world, for women to do so. I'm not sure whether my mother was disappointed by that fact or even gave it much thought. The prevailing culture held, for better or worse, that women were supposed to sacrifice so their husband and sons could achieve success that would in turn bring glory to the family.

Nowadays that approach hardly seems fair and certainly doesn't comport well with contemporary notions regarding the equality of the sexes. That was a different era, and my mother held up her end of the bargain heroically, devoting herself to her husband and children to an extent that almost defies belief. And for that I am eternally grateful, although I wish she had been afforded the same opportunities that her offspring were lucky enough to have.

My father's academic career got off to a promising start. In 1944, when he was in his early thirties, he became a lecturer in history and philosophy at Amoy University in China's Fujian Province. He was a thoughtful, highly educated man—an intellectual through and through. But he lacked a business background, as well as keen instincts in that realm. Over the years, my parents managed to acquire some land, fishing boats, and other material possessions, but they lost all of it when the Communists seized control of the country. My father assumed we would return to Shantou after this whole Communist episode blew over, but as things turned out, that "episode" did not blow over. We never went back, nor did we ever reclaim our land or boats or anything else of value.

When we arrived in Hong Kong in 1949, my father—like many of the hundreds of thousands of Chinese refugees—did not have a job lined up. He had seven people in our immediate family, including himself, to feed (with three more children soon to come), plus an adopted older sister who helped around the house and eight more dependents from my mother's side of the family—her mother, three brothers, three sisters, and a sister-in-law. That was a lot of mouths to feed, but it was an ineluctable feature, and trapping, of the Chinese system: If you are the leader of the family, you're responsible for supporting everybody in the family. In this case, my father had a large family to try to keep afloat—and very little money to do so. But it's hard to escape this exigency in China: While the youngest are supposed to respect the eldest, the eldest is supposed to take care of them, and "them" can be a sizable bunch.

This was the burden my father faced in Hong Kong when we initially settled in the western farming village of Yuen Long and tried to make a go of it. He put most of his money into running a farm, thinking that would be the best way of providing a livelihood for so many people. While his intentions were admirable, he was a better educator than farmer. The farm failed within two years, meaning that all the money he'd brought from Shantou—his life savings in other words—was virtually gone. We had to take many of our belongings to a pawnshop and still had barely enough money to get by on.

Given that my father was now almost penniless, he could no longer support the extended family. One uncle went back to China; the other two uncles left to find work elsewhere in Hong Kong. My grandmother and aunts, unfortunately, had to move out too, which relieved some of the financial pressure on my parents.

The first place we lived in Yuen Long was a large building inhabited by many families. There was no electricity, so we relied on oil lamps for illumination. Nor did our home have any running water, so we had to go to a nearby stream to get water and take baths. Sometimes the water in the stream was high, sometimes it was low, and sometimes it was too cold to bathe in comfortably; but we had no choice—high or low, warm or cold, the dictates of hygiene took precedence, and we bathed no matter what.

My father lined up some teaching jobs in Kowloon and the city of Hong Kong, both of which were far from our home. He had to get up very early to take a bike taxi to the bus stop and catch a bus from there and then a ferry—a journey that took him at least two hours. Between work and all the hours of commuting, he didn't have much time to spend with us afterwards. On some days, in fact, we didn't see him at all.

Sadly, this was rather typical of father's life in Hong Kong. Although he was a highly regarded academician, he never managed to get a commensurately high-paying position. Because he did not speak English, he could not teach at the British-affiliated schools where better salaries were attainable. Instead, he had to cobble together several jobs, often three at a time, none of which paid well. As a result, he spent long hours working and traveling from our home to and between his various jobs, leaving precious few hours in the day for my mother and us.

My mother had long days too, almost oppressively so, typically starting at 5 or 6 a.m. when she made bread or congee (a rice porridge) for our breakfast, assuming we had sufficient provisions even for that. She often did not get to bed until midnight, and sadly, it was not unusual for her to stay up the entire night, attending to various chores she had not found time for earlier. During her waking hours—which, as I said, could be almost endless—she tried to attend to all our needs, making sure that we were fed and clothed, taking care of the house, making our clothes by hand, getting us to school on time, comforting us when we were sick, and helping us with our homework.

On top of all that, she supplemented our income through knitting, embroidery, and other forms of needlework. She knitted sweaters and other goods or sewed flowers onto pillows and bedding—all of which could be sold in town to help support the family. She also made and sold plastic flowers, while fabricating various articles with beads. It was a hard life, which she endured nobly, without ever complaining. But even after pooling her earnings with those of my father, there was still little money to go around. In the morning, we often didn't know whether there'd be anything to eat for dinner.

Mother raised some chickens, though not enough to provide us a steady means of sustenance. Sometimes we secured a bit of food from a nearby church, which served up some Christian gospel while also distributing rice, flour, and other items donated by the United States. We tried other relief agencies and charities when the church's supplies were exhausted. But getting these commodities was by no means a sure thing, given all the poor people living in the area who were similarly in need.

My brothers and sisters and I tried to make the best of it, finding ways to amuse ourselves. Objectively speaking, we grew up in poverty, though not knowing any better, we didn't see it that way. We had a rich, stimulating home life to counterbalance the monetary shortfalls. And we still laughed and clowned around like other children. Apart from donning cheap footwear and clothes that would not win any fashion contests, the most noticeable way in which poverty affected us was in the dearth of food and a gnawing sense of hunger that was usually in the background, though occasionally it leaped to the fore.

Many of our outings, accordingly, involved foraging in the fields around us. Our house was surrounded by farms, and after the crops were harvested, edible things, like sweet potatoes, were left behind, which we gathered. While rummaging through nearby rice patties, we often found water chestnuts, which made for a tasty snack. We also tried to catch frogs because they were fun to play with and, when cooked properly, good to eat, especially the big ones. We also fed frogs to our chickens. One drawback of hanging around rice patties was leeches, which sometimes latched onto our legs and arms. We were afraid of snakes, too, and did our best to avoid them because we couldn't always tell which ones were poisonous.

I started my formal education when I was five, after having taken a test given to everyone planning to attend public schools. A portion of this test contained my first examination in mathematics. Among other tasks, I was asked to count from 1 to 50 and write the results down on paper, in numerical order, of course. Chinese scholars write from right to left, as I'd seen my father do. So I assumed that numbers should be written from right to left as well. This assumption was incorrect. Numbers adhere to the Western convention and are written from left to right. When I wrote the number 13 using my methodology, for example, it came out as 31. In fact, all the two-digit numbers (other than 11, 22, 33, and 44) were reversed owing to my flawed approach. As a result, I failed the exam.

The consequences of this mistake were pronounced: Instead of being admitted to a normal public school where the higher achieving students tended to go, I was sent to a village school, reserved for students for whom expectations were low. Expectations for the school itself were low as well, and the school lived down to its less-than-stellar reputation.

As if that weren't bad enough, we soon moved to a new house located right next to a farm where cow manure was processed into fertilizer. We smelled cow dung most of the time, and when the wind was blowing in just the "right" direction—the wrong direction for us—dried fecal particles sometimes drifted into our residence, a place we affectionately dubbed the "Bullshit House."

To top things off, I now had to walk even farther to reach my substandard school—two miles each way, which is an appreciable distance for a five-year-old of diminutive stature. I had to make this journey alone, often in extreme heat, so my mother gave me an umbrella to block the sun. My short stature, combined with the hemispherical appendage overhead, gave rise to a nickname that I never loved but had to endure because of its pervasive use: "Little Mushroom."

On occasion, said Mushroom would stop for a brief rest at his grandmother's house on his way to or from school, and sometimes she invited him to come the next day for lunch. I started to fantasize about the treats she would lay out before me, but I invariably ended up with something more modest—a small bowl of rice perhaps seasoned with a hint of soy sauce. That gives you an idea of how poor we were—the fact that giving someone a small bowl of rice was considered a big deal. It's no wonder that the kids in my family were often thinking about food. We always looked forward to the New Year's celebration because we hoped to eat better in the next year. In fact, we looked forward to any holiday, when we might partake of a bite or two of chicken or pork, a piece of cake—anything other than the usual staples of plain rice and weak broth.

I was small and thin for my age, the proverbial runt of the litter. Most of the kids who took a similar route to school were bigger and stronger than me and rougher in temperament. They often got into fights among themselves and once tried to blame me for a fight—a particularly nasty brawl in which some of the participants were seriously hurt. My teacher sided with the rough boys and also blamed me. Not knowing what kind of punishment was in store for me, I became so worried that I got sick. My father decided I should stay home for a little while to recover from my ailment (which would now probably be classified as related to stress or anxiety).

I was soon saved by yet another family move. Toward the end of 1954, when I was still five, my father decided to take us to Shatin, which was then a small village due north of Hong Kong. He was slated to begin the next year as a lecturer at Chung Chi College, which had just moved to Shatin, where he would instruct students on various topics, including economics, history, and geography.

At the time, the town's commercial district was tiny, consisting of just a block or two of shops. Shatin now has a population of more than six hundred thousand—a number that continues to climb. Our first house, situated on a hill next to a Buddhist temple, was surrounded by trees, which would have been nice were it not for that fact that they made the inside of the house dark, damp, and gloomy. The walk to the primary school was, again, about two miles. I complained bitterly, insisting that I wouldn't go to school anymore, but my arguments were ignored. However, things changed again after we all got sick during our first year there, each of us succumbing to high fevers and—at least in my case—delirious nights, plagued with bad dreams.

We never figured out the cause of all that illness; maybe it was the clamminess of the house, which made it feel too cool at times and uncomfortably warm at others. In any case, my father decided to move the following year, in 1955, to a nicer house, which we shared with three other families. That building was situated on a hill too, affording great views of the sea, which wasn't too far away. In fact, we could walk down to the sea rather easily for a swim or to collect seashells, starfish, and crabs.

Now that my youngest sister, Shing-Ho, had been born, and my adopted sister Moi Ni had gone off to get married, ten of us were packed into a home with just two bedrooms. That said, it was still the nicest place I lived during my childhood, in part because our next-door neighbors were so friendly and our surroundings so inviting. Tall, flowering trees that bloomed at various times of the year were scattered about; and roses, peonies, and other flowers were planted throughout the yard. We could walk to the sea or hike up into the mountains, or just look out into the distance and admire the vistas. At times like these, our worries vanished. It almost seemed as if we had it made.

Although the house was a great improvement over our previous dwelling, it was far from what you'd called luxurious. It was flimsily built, with walls that were partly composed of mud. During big storms, the whole structure shook, and we feared it would be blown apart. Sections of the house did in fact collapse when major typhoons struck, laying bare the shoddy construction underneath.

And once again, we lived without running water, so we drew water from a nearby stream. Sometimes a selfish neighbor diverted the stream into a reservoir he created in his yard by building a small dam out of rocks and mud, leaving no water for the rest of us. My brothers and sisters and I once banded together to try to clear the obstruction and thereby restore the stream's natural flow. The neighbor, who was a big man, confronted us, but ten other kids from a family whose water supply had been similarly cut off surrounded his house, holding long sticks and demanding justice. Finally, the neighbor relented, and we were again able to fill our buckets in the stream.

That is, until the stream dried up, which happened periodically, forcing us to go to a Taoist temple to get our water. We then had to carry ten-gallon buckets uphill for half a mile, which was a tough job for young kids. When we were little, we put a stick through the bucket handle so we could distribute the weight between two of us. Getting water was always a challenge while we were growing up, making me aware of how it's taken for granted, and often used profligately, in the United States. It's only when you don't have water, or have to struggle to obtain even limited quantities of it, that you can appreciate how important this resource is. One often hears in science class that water is essential for life, and we got repeated reminders of that in our everyday experience.

One nice thing about our forays for water is that it gave us an excuse to go into the mountains. There were streams up above where we'd play on rocks and try to catch fish, which I sometimes kept as pets in a large bowl in the backyard. We'd also try to find nuts in trees (to curb our almost constant hunger) and gather wildflowers, as we could never afford to buy flowers in a store.

My mother had to go to town every day to get food. Sometimes she took us with her, and it could be an interesting scene. In the morning, people lined up to sell their wares on the street. That trade was illegal, and the police periodically moved in to block this commerce. People would run away panicked in all directions as the situation degenerated into chaos. I felt bad that many people, who could not afford to do so, lost their belongings while fleeing from the local constabulary.

We weren't the only family that didn't have much money to buy food. Other poor families would pool their resources, helping one another out during especially lean times. This cooperative practice allowed us to keep food on the table when funds were short, as they often were. In the same spirit, my mother and father went out of their way to help friends and relatives in need, even when we were hard-pressed to meet our own needs. My parents were constantly giving in every way possible, setting an example of generosity and virtue that has stayed with me.

Amidst the daily struggle to get by, we always looked forward to holidays—times when we could temporarily cast aside our pressing concerns and rejoice in the moment. For example, we made a big deal of the Chinese New Year celebration, which took place early in 1956, just as it did every year. Even though we were poor, my mother prepared for an entire month, making her own wine, a special New Year's cake, plus rice cakes and other treats as gifts for family and friends.

The day before the New Year takes on heightened significance in China. Our family, like others, got together for a big meal. My father placed pictures of my grandmother, grandfather, and other relatives on the table and lit some incense, telling us the story of where our ancestors came from. We showed respect to our ancestors by bowing down before the photos a customary three times.

The next day, we started things off by lighting firecrackers. I was usually the one who set off the pyrotechnics. Afterwards, my parents asked all the kids to stand together; and we bowed to them, saying "Happy New Year" and other nice words. My mother gave each of us a small amount of money, typically 1 Hong Kong dollar, enclosed within a red envelope—because red signifies good luck. (That was a modest sum, which was then worth about 15 U.S. cents, but still enough to buy a bowl of noodles.) New Year's was so important to my parents that they sometimes borrowed money in order to offer us even these modest cash awards.

Father then took us by bus to meet his friends and relatives. If we visited a rich friend and received another red envelope, we gave the money to our mother. Through this tradition, I got to meet many people who were close to my father. At these gatherings, the kids sometimes came together to play poker, which we hardly ever did except during holidays.

Another celebration came around in September and sometimes October—the Mid-Autumn Moon Festival. My mother would make a Moon Cake, which is like a cookie with different fillings. Then the kids would run around late at night in the mountains and hills, carrying lanterns. It was dangerous, as the lanterns had a tendency to catch on fire, but it was also a lot of fun.

Looking back at these and other festivals, I can see that even during the most trying times, the hardships and privations of our lives were usually punctuated by some moments of levity.

Once a week, my father taught Chinese calligraphy and poetry to my brothers and me and some boys who lived near us. My father believed that any self-respecting scholar had to be good at calligraphy—a custom that dated back to ancient times. We had to memorize the works of famous poets of the past and then write them down on cheap paper. Respectable scholars, my father taught us, made ink from charcoal ground on a stone, so that's what we did too. It was an arduous task, but our homemade ink was better than what we could purchase in a store.

The next step was even harder: My father asked us to memorize long poems and recite them in front of him. We needed to pronounce the words correctly and speak them forcefully, he asserted, saying, "You cannot get the feeling of a poem unless you read it out loud."

One neighbor complained about the noise from all those kids spouting poetry—though better loud poetry recitations than loud, raucous parties. Sometimes the assignments my father handed out were too difficult for me, but I still learned a lot about Chinese literature and history through these exercises.

I wasn't applying myself in school those days, but I took my father's lessons seriously. He was my most important teacher back then and remains so to this day. This early training with him sparked an interest in Chinese history, literature, and poetry that has never left me. It has even influenced my work in mathematics—not the actual mechanics of solving a problem but more the way I approach a problem, always trying to understand its historical context. Knowing what came before, I've found, can often provide clues as to what the next steps ought to be.

In a more general sense, I benefited from the high expectations my father held for me, even though I did not know how to fulfill them when I was young and, unfortunately, did not figure that out until after he was gone. In addition to the lessons I had with my father, as well as our more casual interactions, I enjoyed sitting in on the animated discussions he had with his college-aged students, who often visited our house. Sometimes they talked about philosophy, entertaining notions that went far beyond the grasp of a young child, but I could still feel the excitement of their words and in this way learned about the power of ideas to captivate people.

That was part of my informal training, which in many ways took precedence over my formal schooling. Now that we had moved to Shatin, I was starting all over again in a new school with new teachers and classmates, after having been bullied in first grade in Yuen Long and missing almost half a year. Sometimes my classmates laughed at me because of my flimsy shoes and handmade clothes, though the derision was never too extreme. Besides, I never much cared about fashion.

One noticeable change was that the classes at my new school were more rigorous than I was used to, especially compared with the remedial institution I had attended, or partially attended, the year before. In second grade, I started to get a sense of what real learning meant, and frankly, I didn't do too well. Nor did I do well in third grade. In fact, I was barely hanging on. The hour-long walk to school and home again was already hard enough and was sometimes too much for me. And I still couldn't shake the Little Mushroom moniker, which I was never fond of.

Sometimes during the return trip, I got frustrated and sat down on the side of the road. My father occasionally sent my third-oldest sister, Shing-Yue, to help me get home. Walking to and from school wasn't the only thing I struggled with. I also didn't do well in the sports that started before school. I wasn't good enough to join in basketball games and was usually too small to take part in the other matches the children engaged in.

While they played their sports, I often wandered around the schoolyard, which was located on a small hill. Once during these meanderings, I came across a human skull and skeletal remains—a consequence of erosion plus the fact that the school was built on former burial grounds.

The school's only bathroom was a six- or seven-minute walk away. Some adults hid there and smoked opium. We therefore kept our trips to the lavatory to a minimum, as we had to encounter these characters practically every time we used the facilities—though perhaps these outings conveyed a hidden educational message, something along the lines of, "Kids, don't do drugs (or you may end up spending more time than you ever imagined in public restrooms)."

At the end of the second semester, Shing-Yue ran into a friend and me after school and asked us how we had done during the school term. I was too embarrassed to reply, knowing my performance had been subpar. But my friend told her I had done great.

"How great?" she asked.

"Oh, he was thirty-sixth in the class!" boasted my friend, who was himself rated about fortieth out of probably not many more than forty.

By fourth grade I started to do better, and fifth grade better still, ranking second in the class, which made my father happy. I did pretty well in math, too, even though there is not much math to be done at that level. We also started to learn English in fifth grade. I hadn't heard or spoken a word of English before, yet something had already happened involving this language that's had a long-standing impact on my life. Hong Kong back then was still under British rule. Because my school received half of its support from the government, every student had to register with the government. But the forms we had to fill out were in English. Since I didn't know any English, my teacher filled out the forms for me. In Mandarin, my last name translates to Chiu, which is the name my father used. But my teacher translated my name from Cantonese into English, which is how I ended up with the name Yau, and I have been called that ever since. (Both Chiu and Yau, depending on which translation you use, are the first name of the famous Confucius, who was born in 551 BCE. A prolific writer and thinker, he stressed that true understanding comes only through hard study. Our father introduced those teachings to us at an early age, so we not only shared a name with Confucius but also learned something about his ideas.)

Years later, both of my sons elected to be called Chiu as a way of honoring the family tradition. But at that time, my father did not care that my primary school called me Yau. I was young then and didn't care either. We never figured that my English name would matter beyond the forms it had been written on. No one could have predicted at the time that I would eventually settle in America and forever be known as Yau.

I didn't learn much English in the fifth grade, but along with my fellow students I got a rude awakening in the sixth grade. We then had a new teacher named Ma who had just graduated from Hong Kong University. Ma decided that the whole class should speak only English—a prospect that scared all of us because our prior exposure to that language had been extremely limited. For the first two weeks, no one understood a word he said. "Do you understand?" he would ask in English, repeatedly and aggressively, yet very few of us even understood the question. For some of the students this was a disaster, as Ma was very strict and handed out bad grades with little hesitation. Some of the kids got so angry that they brought knives to school one day. After classes were over, they surrounded Ma as he was walking to the bus stop and beat him severely, which shows how rough my classmates were. It was a horrible incident that likely gave Ma reason to reconsider his pedagogical practice.

The other big thing that happened in sixth grade was a public exam, administered in every school, that all students had to take to determine which middle school they would go to. This exam was very important for everybody and was made even more important by the fact that in Hong Kong, middle school and high school are combined. Preparing for that exam, accordingly, became our sole priority during the first part of the year—or at least it was supposed to be. Our class of about forty-five students was divided into seven study groups. Because I had ranked second in my class the previous year, I was picked to lead one of the study groups composed of about seven boys. Of course, I was just a kid myself, a mere eleven years old, and was in no way qualified to oversee the studies of my oftentimes rambunctious peers, nor did I consider myself up to the task of keeping them in line.

On the first day of our "school without walls," I left home at the usual time to meet with my classmates, but we didn't know what to do. We didn't have any books, nor was there a public library where we could study. After a few uneventful get-togethers, two members of our group went their separate ways, but four of the boys stuck with me. Not knowing how we could productively make use of the time, we wandered around the Shatin area instead, and I thus began my brief stint as a juvenile delinquent.

Our activities during this period were sometimes benign but other times less than honorable. We hung around the marketplace, occasionally stealing things when the opportunity arose. Sometimes we met up with other "gangs"—encounters that were not always amicable. We once ran into a group of rough kids near the railroad tracks. After sizing up the situation and concluding that we were seriously overmatched, I decided to take the offensive. I grabbed some rocks and started hurling them at the other kids, which, to my surprise, scared them off. My mates viewed my initiative as a sign of bravery, helping to reinforce the sense among our crew that I was a capable leader. I'm not particularly proud of this moment, but it did show me that even though I was physically smaller and weaker than the other kids, I was not afraid to stand up to the band of ruffians (not so dissimilar from ourselves) that we then faced. This strength of spirit served me well in other difficult situations in the future—even those arising in mathematical or more generally academic settings that have nothing to do with gangs of errant youth and in which the typical weapons of choice assume subtler, more devious forms than sticks and stones.

As for my own group of wayward sixth-graders, we got into fistfights on other occasions but also engaged in more innocent pursuits—playing marbles, hanging out at the beach, or heading into the mountains to catch birds or snakes. But the fact of the matter is that for half a year we mostly wandered around, not doing much that could be called useful and surely doing nothing that would qualify as study or would advance our academic ambitions.

During this time I left home every morning at 7:30 a.m. and returned every afternoon around 5 p.m., the same hours I kept on regular school days, so my parents (and siblings) had no idea that any of these extracurricular activities were going on. But the day of reckoning arrived soon enough in the spring when we all had to take the middle school exam and practically everyone in my group failed. Before the end of the school year, the Hong Kong government announced who had passed, printing the names of those students in the newspaper. That afternoon, when I was having a grand time playing with some kids in our neighborhood, one of my sisters interrupted. "Father wants to talk to you," she said solemnly.

I found my father inside our house, and he was very angry, as angry as I'd ever seen him, because my name was not on the list. "You are doomed!" he said. That summed things up pretty well, although there was still a glimmer of hope. I noticed that on the next page of the newspaper my name appeared on a waiting list, of sorts, for students who had not secured a spot in a public middle school but were still eligible to apply to a private school.

My father had been ready to punish me severely but was relieved to see that there was still a chance for me—to see that all his efforts to teach me classic Chinese poems and historical tales had not been wasted. Fortunately, he was familiar with a good private school close to our home that was arguably the best in all of Hong Kong, the Pui Ching Middle School, and he knew some people there—people in high places. The president of this school respected my father and had once offered him a job. My father was also on good terms with the chief secretary, another powerful figure at the school. I don't know how those connections factored into the equation, but I was given another shot. The school asked me to take a separate entrance exam, and this time I worked very hard to prepare, realizing it might be my last chance to make something of myself. I'm thankful that I did not squander this opportunity, doing well enough on the exam to gain admittance to Pui Ching.

Better yet, the government of Hong Kong was going to pay for my education there, which we could not possibly afford. The only catch was that this financial aid money is typically given at the end of the school year, meaning that we weren't able to pay our bill at the beginning of the term. Each year, I had to go to the school's president and ask whether we could pay the money later, after getting support from the government. It was somewhat embarrassing to have to make this request at the start of every school year, but everything worked out in the end.

I was truly lucky to be at Pui Ching because it was a first-class institution. Daniel Tsui, a graduate of the school who was ten years older than me, won the Nobel Prize in Physics in 1998. Eight graduates of the school are now members of the U.S. National Academy of Sciences, and three of them (including me) won the National Medal of Science. Yum-Tong Siu, a colleague of mine at Harvard and an accomplished mathematician in his own right, is another illustrious Pui Ching graduate. My former middle school classmate Shiu-Yuen Cheng has served as the chair of the mathematics departments at both Hong Kong University and the Chinese University of Hong Kong, among other posts.

The point I'm making is that Pui Ching is a fine school, and going there turned things around for me. I got in by a stroke of luck, for had I done well on the middle school exam the first time around, and not goofed off for six months beforehand, I would have attended a public school that was not nearly as good. That happened to my older brother Shing-Yuk—a good student who did not choose to spend the better part of sixth grade as a vagrant and general layabout. All I can say in my defense is that I recognized how fortunate I had been and was determined to make amends.

I spent six years at Pui Ching, which encompassed both middle school and high school. We spoke Cantonese in our classes, but most of our books were in English, except for those on geography and Chinese literature and history. The only class that was not taught in Cantonese was our English class. In that course, we had to do our homework in English too, so we became fairly used to the language by the time we graduated.

The math teachers at Pui Ching really stood out. For the most part, they were excellent and got me to pay more attention to this subject than I ever had before. The physics teachers weren't so great, which may be one of the reasons I didn't become a physicist. The chemistry teachers, on the other hand, were outstanding, though the subject matter never captivated me. I didn't love mathematics at first either, but the more time I spent with it, the more intrigued I became—a development my father supported wholeheartedly. Because he was a philosopher, he helped me perceive the world through a more abstract lens. Logic lies at the heart of both mathematics and philosophy, which is one of the reasons my father appreciated math so much. He was glad to see my growing interest in the subject. But beyond that, he was always encouraging his children to find the thing, whatever it might be, that really excited them.

At Pui Ching, my fellow students and I often heard about Yum-Tong Siu, who was six years older than me and already a legend at our school, praised for his mathematical prowess. He got the highest grades and test scores of anyone in Hong Kong and was well known for that reason. Our paths crossed many years later in the United States, and at times relations between us became tense. But that wasn't the case when I was in seventh grade and life was much simpler and interpersonal dealings more transparent.

The Pui Ching School, which was founded in 1889, is located in Ho Man Tin, which was then a small town in the western part of Kowloon and has since become much more urbanized. Getting there wasn't so difficult, and thankfully, no one was calling me Little Mushroom anymore. I left home every morning around 7:15 to walk to the train station, where I met up with several friends. It took only about fifteen minutes to reach Kowloon by train and another fifteen or so minutes to walk from the station to school.

Pui Ching was owned and run by the Baptist church, and the principal was a powerful figure in the church, although I never got wrapped up in the religious side of things. School started at 8:30. We typically had a few classes in the morning, lunch at noon, and a couple of more classes in the afternoon, which ended at 3:15. The train left at 3:30 sharp, so we had to run from school or we would not catch it.

I had some difficulty adjusting to the academically rigorous climate at Pui Ching, as my previous school in Shatin had been populated mainly with farm kids who had more casual attitudes toward academic achievement. This school was more upper crust, and I was teased for wearing shabby clothes and bringing scraps of leftovers for lunch rather than buying a more respectable meal from one of the nearby eateries.

My middle school teacher was not entirely pleased with me, owing to my tendency to chat too much during class. The school ran on a quarter system, and at the end of each quarter students were required to show their parents comments from teachers and get their signature. "He likes to talk and likes to move around," was the first comment about me. The comment about me for the second quarter was similar, whereas I "showed slight improvement" during the third quarter.

I worked pretty hard during my first year, much harder than I was accustomed to, but evidently not hard enough according to the teachers' reports. Two subjects gave me the most trouble: music and physical education. I sang miserably in class and was always out of tune. My teacher accentuated this problem by routinely picking the students who sang best and worst and asking them to sing in front of the class. My solos were tough to endure for everyone, especially for me, and no one wanted to sing with me for fear that their own grades would suffer—a kind of guilt by association.

My teacher was not charitable about my musical shortcomings. Back then, my hair had a tendency to stick up, and try as I might, I could not get it to stay down. "You can see how lazy this kid is," my music teacher would often complain. "He cannot sing and does not even bother to comb his hair."

I failed music the first year even though I made a concerted effort to learn how to sing; I practiced every Saturday with my cousin, who was a piano teacher. I retook the music exam in the summer, and this time I passed. But there is still a red mark in my permanent record—a mark signifying failure that no student wants to have.

I also got a red mark (or two) in physical education. It took me about nine and a half seconds to run fifty meters, which is considered slow, and I could do only two pull-ups. I managed to do thirty sit-ups, though that still fell short of the fifty we were supposed to do. My results weren't anything to brag about, but I tried, and that ought to be worth something.

Mathematics didn't excite me during the first year, probably because it wasn't presented in a challenging way. Our teacher was barely twenty years old, and she acted more like a big sister than a teacher. Owing to her inexperience, she had trouble bringing the subject to life. However, when I became a teacher several years later, and struggled at first, I was more sympathetic toward her situation.

During my second year of middle school, I got a taste of what mathematics was really like. Our teacher was very strong, and he taught us Euclidean geometry. I was amazed to see how far one could go, and how many theorems one could prove, starting from five simple axioms. For some reason, which I couldn't quite put into words at the time, that idea made me happy, and I started toying around with this approach on my own.

I posed the following question, which I assumed was my own invention: Using just a ruler and a compass, can you construct a unique triangle if you know any three of the following quantities—either the length of the triangle's sides, the size of its angles, the length of the median (or line segment) that runs from the midpoint to the opposite vertex, or the length of an angle bisector? *And was this statement always true?* I realized from the outset that there had to be at least one exception: If all you knew was the size of the three angles, that alone would not define a unique triangle. For there could be an infinite number of triangles of different sizes, each having those three angles. So the statement clearly did not hold in that situation.

All the other possibilities checked out, as best I could tell, except for one, which held my interest for quite some time. Suppose you know the length of one side of a triangle, one angle, and the length of one angle bisector. Can you construct the corresponding triangle, using just a compass and ruler? I worked on this problem for the better part of a year and made little headway. I thought about it when I was walking to the school and riding in the train, but I could not verify that it was true. Although this was frustrating on one level, it was also intriguing, because I was eager to find out whether the general rule I had stated broke down in this case.

Some of my classmates were pretty rough kids who would intimidate me at lunch or during outdoor sporting contests. One corpulent lad, for example, had the nasty habit of squeezing my forearm so hard that it tingled and then started to lose all sensation. The boy's motivations for doing this were not always apparent, but his fingers made an impression on me. About the only thing that saved me from this kind of unwanted attention was that he and the other kids knew I could help them with math. So they tried to stay on my good side.

Once, while playing soccer, I got hit in the face with a ball so hard that

I almost passed out. The other boys found this endlessly amusing. They made fun of me for this and on many other occasions. At one point, I got so annoyed that I said, "If you're so great, here's a problem I came up with on my own. Let's see if you can solve it." I repeated the problem about the triangle that I'd been grappling with, and of course, no one in the group could get anywhere with it. My math teacher, upon hearing of the problem, could not solve it either.

Our school met Monday through Friday and on Saturday until noon. After class on Saturdays, I had a little time before the train came, which I often spent in a bookstore in Kowloon. I read math books there since I couldn't afford to buy any. One day, I found a book that discussed the same problem I'd been working on—the problem I thought I had invented. I learned that it could not be solved, which came as quite a relief. The book cited a recent argument that proved you could not construct one, and only one, triangle that satisfied three of those conditions.

I was excited to see that "my problem" had stumped other people and was only recently shown to be insoluble. I further realized that this same problem was similar to one that dated back many centuries: Could you trisect an angle if you had only a ruler and compass? No, you could not. Nor could you solve another long-standing problem, "squaring a circle," which means finding a square with the same area as a given circle, again relying on only those two implements. I was proud to find out that my problem was in the same category as these two classic problems. The fact that I could not solve it was not a sign of defeat. On the contrary, it put me in good company.

Which is a long-winded way of saying that I enjoyed math during my second year at Pui Ching, and did pretty well at it, though I continued to fail in music and struggled with English too. Our Chinese literature teacher, Miss Poon, was a young woman about twenty-two years old. She was very stern and gave all the students a hard time. I still recall the glasses she wore, which were extremely pointy. She no longer had those glasses when I ran into her many years later. I asked her why she had adopted such a severe style, and she said that our school was famous for having kids who were naughty, especially the boys. She thought she might intimidate unruly kids by having such sharp corners on her eyewear.

That year the president of the school addressed the entire student body at a big assembly. When he came to the podium, the kids were making so much noise he could not speak. He chided us for our disrespectful behavior and admonished the teachers, instructing them to control their students. Many students, he added, were not wearing a tie, in defiance of the school's time-honored tradition. Which was true of me at that moment—a fact that did not go unnoticed by my teacher and "classmaster," Miss Poon. I was wearing the standard uniform but not a tie, though I had a somewhat legitimate excuse: My windpipe (or trachea) was just one-fourth the normal diameter. Wearing a tie made it harder for me to breathe, so I normally did not put it on until just before school started. On that day, however, the train had been late. I ran from the station with the tie in my pocket and had not had a chance before the assembly to put it on. I started doing so after the president chastised us, but by then it was too late.

Miss Poon ordered me to speak with her after school. During this meeting, she told me I would be punished for my dress code violation. You got points for every infraction, and I would get two points for not wearing a tie, which she considered a big insult to the president and an affront to the whole learning community. If you got nine points, you were permanently kicked out of the school. She also would be sending a report to my father about this incident, which I knew would not go over well. I was in tears over my impending punishment, not knowing what it might lead to.

Miss Poon looked at me as I sobbed, as if she'd never really noticed me before. While I awaited my "sentence," she caught me off guard by asking why I was wearing such skimpy clothing. I told her that was all I had. She then wondered, perhaps owing to my thin frame and pale complexion, whether I might be malnourished, asking what I'd been eating. When I told her what, and how much, I consumed in a typical day, she said, "Your father is a professor, yet you don't have enough to wear or enough to eat?" After I explained our circumstances in fuller detail, she showed great sympathy, and even gave me some milk powder and other food to supplement my meager diet.

This incident was something of a turning point for me. I was touched by her kindness, which, in all the years I could remember, had rarely come from a teacher or school administrator. I decided that I would not let her down. Henceforth, I vowed to become a better student, which I did—a development that pleased my father to no end. My second year of studies went well from that point on. In addition to becoming excited about math, I also began to learn some elementary physics.

Despite my new resolve, my third year at Pui Ching turned out to be a disaster for reasons that were beyond my control. My second oldest sister, Shing-Hu, who'd been at high school in Macau, came home with a serious illness. My mother dropped everything to take care of her, but unfortunately Shing-Hu got sicker and sicker and died at the age of nineteen in September 1962, just after my school year had begun. It was a tragic turn of events that left us all shocked and profoundly sad. I saw my father cry for the first time in my life, which was jarring. I simultaneously felt a sense of loss I had never experienced before.

But that was just the beginning of our troubles. At the time, my father headed the programs in philosophy, Chinese history, and literature at (the now defunct) Hong Kong College—a school he had founded with a man named Shu Kui Chan who was serving as the college's president. Father's career seemed to be going well: He had just finished writing a book on Western philosophy and was about to start another on Chinese philosophy. But various complications ensued, in part because Hong Kong was a very complicated place. It was then populated with a large number of refugees, including our family, and a fair number of spies from mainland China, Taiwan, the United States, and the United Kingdom. The Taiwanese government, my father recounted, made specific overtures to the leaders of Hong Kong College. Taiwanese officials told people like Chan that after they overtook China, a prospect they considered inevitable, he would be awarded a cushy government post, such as mayor of a mainland city, provided that he allowed spies to routinely infiltrate the college.

While father was strongly opposed to the offer, he told me that Chan appeared to have no problem with it. Chan tried to get rid of my father so that someone more sympathetic to the Taiwanese proposition could take his place. The terms of my father's contract would have prevented Chan from firing him outright, but my father decided to resign in protest, as he no longer respected the values of our school's leader.

My father left his job in November 1962. At roughly the same time, he lost his teaching position at Chung Chi College owing to the close relationship he had with the college head, Dao-Yang Ling, who was on the verge of being ousted from his leadership post. This turn of events caused our family's income to drop dramatically. The career setbacks to my father, coupled with the recent death of his daughter, sent him into a deep depression.

About two months later, during the Chinese New Year celebration, father got sick, feeling so uncomfortable he was unable to sleep at night. We thought his illness stemmed from a serving of crab he had eaten that might have gone bad. While that could have been a contributing factor, there turned out to be a more serious underlying condition that had yet to be diagnosed. As we had very little money at the time, he tried to treat himself with inexpensive Chinese herbs and medicines. Nothing worked, and his health continued to decline. My mother sought help from her younger brother, who had become rich from running a successful private Catholic high school—the same brother my father had generously supported years before. My mother tried to borrow money from him so that we could obtain better medical treatment, but he refused to lend her any.

My mother was proud and hated to beg, but she asked for assistance from all quarters, desperate to help her husband. In April 1963, some students of my father came through, pitching in to send him to a hospital where Western practitioners could treat him. We soon learned that he was suffering from urine poisoning brought on by kidney cancer. My father went to a hospital for treatment, though we couldn't afford it. Within a couple of weeks, he could no longer talk. It broke my heart to see such a wise, eloquent man unable to speak.

I visited him often, even though it was a long, difficult trip, involving many transfers, to reach the hospital from Pui Ching. When my father's situation became dire, one of his students arranged for us to stay in a hotel near the hospital, so we didn't have to travel so far to see him. That was the first time we'd ever been in a hotel though we were surely not celebrating. One night in June, after stopping off at the hotel, I returned to the hospital to find my mother in tears. I didn't need to ask her what had happened—one look at her face told me all I needed to know.

My wonderful, amazing father—a noble scholar who placed learning and honor above everything else—had just died. The entire family was devastated. It was as if an earthquake had torn through our home, ripping the foundation apart, flattening the upper stories, and leaving the rest in rubble. Everything had suddenly and irrevocably changed—and, it goes without saying, changed for the worse. Life as my family had known it had come to a harsh end, and we had no idea what would follow.

CHAPTER TWO

Life Goes On

MY FATHER'S DEATH hit me hard, throwing me into an unfamiliar state in which I felt a weird mixture of things, all unpleasant, all at the same time. A powerful sadness welled up in me from a deep place I'd never accessed before. I felt a dull ache that I couldn't localize, as well as an all-embracing numbness.

That was on the physiological level. But I also felt as if I'd lost my moral compass, because my father was a righteous man who was always pointing us in the right direction, always teaching us about the importance of hard work and positive values—lessons that were often inspired by the writings of Confucius. With my father gone, it seemed as if the center of gravity, the organizing principle in our lives, was gone too.

Our circumstances were so pressing, however, that there was no time to wallow in grief, nor was there any real opportunity for denial to take hold. I realized that not only had everything about my family's situation changed but that I, personally, would have to change too. Right away I felt pressure to start earning some money to support the family. But it was more than that. Without my father to lean on, I knew I had to grow up fast and start making decisions for myself—decisions that would affect the rest of my clan as well.

His death was therefore a turning point for me. I was forced to abandon the Chinese notion, long drummed into our heads, that we could always count on a strong family leader—someone who'd always be ready and waiting to take care of us. The moment had arrived for me to stand up for my own future. While I was doing that, moving on as best I could, I still had a gnawing desire to make my father proud, even though he was not around to see it. His confidence in me had been unwavering throughout our fourteen years together, but I did not always measure up to his expectations.

To my surprise, I started, quite spontaneously, to recite some of the Chinese poems he had introduced me to years earlier—as a way to feel more connected to him. I used to look at those poems in a half-hearted manner, only when asked, but I now took them more seriously and memorized them, just as Father had instructed us to do. Reciting those poems became not only a hobby for me, but also provided some relief for my sorrow, as well as helping me get through other tough times yet to come.

In addition, I began to read some of the philosophical books in my father's collection, which were by no means easy to comprehend. Furthering my education was not my primary motivation, although that happened as a matter of course. Instead, I was aiming to get a better grasp of what my father liked to think about. In these texts, I found traces of him—threads that triggered memories that in turn had a calming effect on me. I came to these exercises naturally, almost subconsciously. They helped strengthen my ties to my father, even after he had left us.

I had a new attitude toward school too, resolving to try harder and become more focused than I had been in my previously carefree existence. The stakes were higher now, and I did not want to let him or my mother, or even myself, down. Achieving excellence in schoolwork, as far as I could see, offered me the only conceivable pathway to success. I'd have one chance to distinguish myself, and if I failed, there would be nothing to fall back on.

Between the loss of my father's income and the medical expenses that had accrued over the several months before his death, our savings had been exhausted. There are no social security payments, retirement benefits, or pensions in China. All you have is your salary, and when a job ends—or worse, when an employee dies—usually not much is left. In our case, nothing was left, and we owed half a year's rent, along with a stack of unpaid bills.

But our first order of business was the funeral. Showing respect for the deceased is a big deal in China. The ceremony we were preparing was intended not only to honor my father in the best way possible, but also to preserve the dignity of our family. My brothers and sisters and I had missed school for several weeks before and after my father's death. My sister Shing-Yue and students of my father took care of the funeral arrangements, while the rest of us helped out as best we could. First, we needed to find some land on which to bury him. That cost money, of course, and we had to pay the funeral home too. Luckily, my father's friends, who were reasonably well off, defrayed some of these expenses. That enabled us to purchase a small burial plot in the New Territories region north of Kowloon.

My siblings and I didn't know much about funerals and basically did what we were told. One of the things we were told to do was to spend the night before the ceremony in the funeral home, which, according to tradition, is supposed to protect the good spirits, or maybe fend off the bad spirits. We were obedient, even though we didn't know what we were trying to accomplish with regard to the spirits, good, bad, or otherwise. I read all of the poems displayed in the funeral home that my father's students had written. These poems assumed a particular form called a "pairlet," consisting of two sentences that are related to each other. I enjoyed reading them because I learned things about my father, and about how other people viewed him, that I hadn't known before.

The next day, in keeping with tradition, we all dressed in white and knelt around a picture of my father that was surrounded by flowers. Whenever somebody came to pay their respects, he or she would bow three times and we would bow too. This went on all day. It was a draining experience but also very moving. Although I was filled with sorrow, for some reason I did not, or could not, cry.

Afterwards, there was much to do. We had to face our many problems, including the rent, which was many months overdue. Fortunately, our landlord was compassionate, and knowing that we were destitute, told us we wouldn't have to pay back the money if we left soon. Our mother found a cheaper place in Shatin for us to live, which was nothing like the nice house, with ocean views, that we had occupied for several years. In fact, it was a two-room shack located next to a pig pen. It's no surprise that living next to a pigsty can be smelly, but it can also be quite noisy. Our next-door "neighbors" got started early in the day, before 6:00 every morning, eager to begin the grunting, snorting, wallowing, and general cavorting that pigs are famous for.

Needless to say, this was not the most serene spot in which to live, but the price was right—or at least close to being affordable. Seven of us—down from ten following the deaths of my father and Shing-Hu and the departure of my oldest sister, Shing-Shan, to England, where she was training to become a nurse—were crammed into this tiny shack. The accommodations were about as crude as you could get, and other kids in the neighborhood looked down on us for being so poor and living in such a miserable excuse for a house.

Of course, that was nothing new; we were used to derision of that sort and the attitudes that go with it. But there was no denying that we were at a low point, surely the lowest we'd ever been. We all hoped that we had "bottomed out" and that things would soon begin an upward turn.

That's when my uncle jumped in, offering us a way out of our predicament: He would get a farm somewhere near Hong Kong, he said. We could then quit school and work for him, taking part in the proud tradition of duck husbandry. While some might consider this a generous offer, it sounded like a nightmare to me. Fortunately, my mother agreed and would have none of it. Even in our precarious position, she knew that accepting such an offer was beneath our dignity. Instead, she wanted us to live according to our father's wishes, which was for us to continue our education and become scholars, or at least to go as far along that road as we could. She felt, as had my father, that gaining knowledge and cultivating the mind was more important than pursuing money. There had to be more to life, Father had told us, than simply attending to one's material needs.

It was a struggle, given our lack of resources, but my mother somehow found the energy to pay the necessary fees so that we could stay in school. This surprised many people, even some of our teachers, who had expected us to drop out at any moment. Mother was malnourished and anemic at the time, as she had been for years, yet she did everything possible to keep us from suffering from dietary deficiencies. Sometimes, when we studied late at night and our energy was waning, she would serve us flavorful broths of beef liver or pig brain, which never failed to give us a boost and temporarily raise our spirits.

Looking back on all that my mother did, I'm amazed at the strength and determination she displayed under such duress. Some people have said that I can be incredibly persistent and stubborn—traits I have applied, for instance, toward solving difficult math problems—and I believe I inherited some of that resolve from my mother. Her encouragement, even while she was suffering such hardships, motivated me to pour myself into my studies. And when I eventually became known in the academic world, she was grateful that her efforts had been rewarded.

I was grateful too, as a fourteen-year-old, when she declined her brother's attempt to recruit us into the duck trade, which would have consigned us to a life of drudgery. Her decision conformed not only to my father's wishes, but to mine as well, because I was already committed—to the extent one can be at the age of fourteen—to making my mark in academia.

My first order of business on that front was to take all the exams I had missed during my weeks-long absence and prepare for the big final exam coming up near the end of the third year. I did well in the math portion of that exam, as usual, and in most subjects except for physical education, as usual, too.

The journey to school from our current dwelling, the "Pig House," was longer than before, because I had to walk nearly an hour to reach the train station. That made for a lengthy round trip, leaving little time for my schoolwork or for sleep. A former student of my father's, K. Y. Lee, offered to help. An elementary school had just been started on the top floor of a new seven-story building the government had erected in the wake of a typhoon that had killed scores of people and destroyed many buildings. The new school was located closer to Pui Ching, and Lee told me I could stay in the classroom at night in order to shorten my commute.

I stayed there for more than a year, helping out the young students in my spare time. Like me, these children came from poor families, and most of them were delightful to be around. The accommodations, however, fell something short of spartan. There was no bed, so I usually slept on a table that was about two feet wide and five feet long. Fortunately, I was not very tall in those days, but I did fall off the table from time to time because it was so narrow. There was one bathroom on that floor, and it was barely tolerable from a sanitary and olfactory standpoint. There were also shops and food stands on the first floor where I could get a (very) basic meal of noodles or rice for 1 Hong Kong dollar.

Former students of my father who lived in the building sometimes hung out in the classroom at night, and we'd talk or play chess. But they never stayed very late, which left me alone for many hours to read and get my own coursework done. If I ever overslept, the younger school kids would be sure to wake me up, with not-too-delicate nudges or pokes, when they arrived in the morning. It was a lonely existence, especially compared with the high-density quarters I had been accustomed to. I went home about once every two weeks to visit the family and wash my clothes, but other than that I learned how to survive on my own, and that's always worth knowing.

Still, I needed to bring some money in, both for my personal expenses and for the others back in Shatin. Shing-Yue had already taken a job as a primary school teacher, giving up her chance to study at a university, in order to generate earnings for the family. My eldest sister, Shing-Shan, began sending money from the United Kingdom once she learned of our father's death. It was clearly time for me to step up too.

That's what prompted me in 1964 to start tutoring in mathematics—a small step at the time that nevertheless helped launch me on my current career path. I was about fifteen when I got that going, working with kids who weren't much younger than me. It was hard to get started in this line of work because I didn't know how to find pupils, nor did I have a telephone to make it easy for prospective clients to contact me. Fortunately, one of my fellow students at Pui Ching, Ying-Cai Zeng, thought that tutoring sounded like fun. He had a telephone at home and placed an ad in the local newspaper, even though, as things turned out, he never did any tutoring himself.

In this way we lined up the first client, a student at a prominent high school who was just a grade behind me. I earned 25 Hong Kong dollars for a month's effort, which was almost enough to cover the cost of my meals. That was a start. My mother then found several more pupils through a government agency, which made me happy because I'd have more money to give her. One student was a sixth-grade girl, a few years younger than me, who'd been failing math. She was struggling with simple arithmetic problems like the following: "If you go to the farm and see thirty-six chicken legs, twenty-eight cow legs, and sixteen horse legs, how many animals have you seen?" She was asked to memorize a formula for solving this problem, but I offered her a whole new way of approaching problems like this and other problems as well. Her mother was nervous because I was teaching her methods for solving equations that went well beyond the sixth-grade level. But the strategy quickly paid off. Within a month, she was scoring 100 on her math tests. Her mother was so ecstatic that she wanted me to teach all her daughters English. I declined that offer because my English was pretty shaky back then—and still is a bit rough around the edges, even after all my years in America.

Tutoring kept me busy, especially when combined with my high school workload. I had moved on to the tenth grade and was doing well enough, not only in math but in Chinese literature and history too, which pleased my mother. Although I was forced into tutoring because I needed the money, I got more out of it than I expected: The process of making math more understandable to kids helped clarify my own thinking on the subject. I found that teaching mathematics could be fulfilling, and that discovery helped push me along the route I've followed ever since.

I got another push when I came across a book by Loo-Keng Hua, one of the most eminent Chinese mathematicians of the twentieth century. This book, which was on number theory, was my first initiation into higher math. It was a revelation to me, and I read several other books by Hua that were also wonderfully written. I saw that mathematics could be a thing of beauty, something to marvel at. This and other inklings I had received over the years—such as my introduction to Euclidean (plane) geometry—helped persuade me that mathematics would be my calling. And it would not be too much of an overstatement to say that coming across Hua's books at that time, in the wake of the despair and aimlessness I had felt after my father's death, gave my life a direction and a sense of purpose that I was suddenly eager to pursue. Of course, I still had a couple of years of high school to get through, and probably some time in college as well, before I could try to put my stamp on the field.

One of the notable aspects of eleventh grade was that I finally started to learn calculus—an elegant set of techniques developed some 350 years earlier by Isaac Newton and Gottfried Leibniz that is still central to much of the current work in mathematics and physics.

By that time, my family had moved from the Pig House to a somewhat nicer dwelling in Shatin, nestled among pine trees with a stream nearby that rushed down from the mountains. We had built this house cheaply, with the help of friends, relatives, neighbors, and relief agencies in the Hong Kong government. My mother's wish of having her own house had finally come true. Not surprisingly, the place was tiny—a bedroom and a living room—barely big enough to fit all seven of us at the same time. And it was, in keeping with our recent history, also quite primitive. We relied on kerosene lamps, as we had no electricity, and a wood fireplace for cooking. Once again, there was no running water on the premises. But there were plenty of snakes in the vicinity, some poisonous, and it was my job to take care of them when they ventured inside.

I moved back to Shatin and slept in the attic, which was only accessible via a ladder. The ceiling was so low I had to crawl to get around and could barely sit up. There were also poisonous spiders and scorpions up there, which caused some concern. But despite all that, I enjoyed being with the family after having spent the previous year sleeping, not so comfortably, in a barren, isolated classroom.

The rural setting of our new home, by contrast, was quite pleasant. My mother planted fruit trees in the yard, and we also kept a few dogs, chickens, and geese—all of which helped liven up the surroundings. The geese in particular were a welcome addition, as they scared away snakes that might otherwise have been inclined to make unannounced visits.

In the eleventh grade I took an important test, the Common Exam, which one needed to pass in order to graduate from high school and go on to college. Fortunately, I did all right on that. During twelfth grade, I lived for several months in the house of a former student of my father's, Pak Win Lee, so that I could tutor his nephew. The house was extremely luxurious, having amenities I'd never been exposed to and in many cases didn't even know existed. They even had servants to attend to their needs. I'm thankful that in my capacity as a live-in tutor, I was treated respectfully. But as someone who grew up in poverty, I was also glad to see that in this house the servants were treated respectfully too. I would have been extremely uncomfortable had that not been the case. Still, the contrast between my plush quarters there and the attic and tabletop where I had previously slept couldn't have been sharper, and this brief episode spoiled me a little. I had gotten a glimpse of how "the other half" lived. Although I didn't need all the frills of the upper-class lifestyle, I saw that life could be better, and easier, if you didn't have to struggle for every scrap of food.

The biggest thing in twelfth grade was the college placement exam. My best friend, Siu-Tat Chui, who was ranked number one in our high school, failed the Chinese literature section of the exam and was not able get into the Chinese University of Hong Kong (CUHK). Tat, as we called him, was one of the most brilliant people I'd ever met, but he was forced to spend an extra year in high school. However, when he graduated, he won big awards in practically every subject, including math. The president of our high school had even spoken up on Tat's behalf, but the head of CUHK still denied him admittance. The system had clearly failed him. Tat was so fed up with Hong Kong that he decided to attend college the next year in Montreal. I also thought about studying abroad, but the application fee was expensive enough that it alone would have placed an unwelcome burden on my family.

I somehow managed to talk my way into the General Certificate of Education (GCE) exam run by the British school system—which was similar to what was then called the Scholastic Aptitude Test (SAT) in the United States—even though I was not eligible to take the GCE because I'd been educated in Chinese schools. I did well on the mathematics and English portions but failed the chemistry section, which had a serious laboratory component. Pui Ching didn't have the facilities for the experiment I was supposed to carry out, so I tried to do it in a friend's basement, using makeshift equipment, and got predictably poor results. As a result, I could not apply to any British schools, though I did pass the placement exam for CUHK—with some difficulty, like my friend Tat—and that's where I went.

Chung Chi College, where my father had taught, was part of CUHK, and that's where I went first so I could stay close to my family. My older brother, Shing-Yuk, was going there too. I didn't apply to study in another country, as some of my peers did, but I wasn't giving up on that idea altogether, as I felt that I would need to go to Europe or North America at some point if I wanted to become a first-rate scientist. I never lost sight of that idea, although I was content to start things off at Chung Chi, which I entered in the fall of 1966.

The chairman of the math department, who was named Tse, was a nice guy, as well as a friend of my father, although he was not the most accomplished mathematician. He gave an introductory speech to the ten or so students majoring in the subject, which he hoped would inspire us. "You've come here to do mathematics," he said. "The sad truth is that you may not be good enough to become a pillar of this hall of mathematics. But if that's the case, you can at least paint the wall." Those words might have sounded depressing to some, but I found them encouraging: He was telling us that we all could make some contribution—in our own ways, large or small—toward progress in the field as a whole.

I soon found that the standard freshman math courses were too easy for me, so I was given permission to skip the classes and just take the exams to demonstrate my proficiency. That enabled me to use the time to take more demanding courses, including linear algebra and advanced calculus. A lecturer named H. L. Chow, who had a master's degree from the Courant Institute at New York University and later earned a PhD in England, taught the latter course.

In Chow's class, I learned about the Dedekind cut, which was invented in the mid-1800s by the German mathematician Richard Dedekind, who was a student of the great Carl Friedrich Gauss and a contemporary of the equally great Bernhard Riemann. Through this technique, Dedekind showed how, starting from integers (consisting of counting numbers like 1, 2, 3; their negative counterparts; and 0), one could construct rational numbers (like 1/2 and 3/4) and irrational numbers (like the square root of 2 or pi, which cannot be expressed as a simple fraction). From there, one could build the real numbers, which encompass all the rational and irrational numbers, thereby covering every single point on a number line, including the integers and everything in between.

I was blown away by the fact that one could take integers, which are familiar to most primary school children, and use them—through a step-by-step procedure—to create something as vast and complicated as the real numbers. It reminded me of the excitement I had felt in eighth grade when I was exposed to plane geometry and saw how far one could go starting from a handful of straightforward axioms. I wrote a letter to my teacher, Chow, expressing my appreciation. "I finally understand why mathematics is so beautiful," I told him. "I am relieved to discover that the subject I love, mathematics, can do what I thought it could do." And I soon learned it could do far more.

I'm not sure what Chow thought of my note, as I don't recall a response, but he probably viewed it favorably. He may have been pleased by my enthusiasm and positive attitude toward the subject he taught, because we soon became friends. He invited me to his house a couple of times, which was very considerate, and his wife was extremely kind. The main problem, for me, was that they had eight cats, and the associated smells pervading their residence were so overwhelming that I almost passed out. It took all of my fortitude to conceal my discomfort and keep from bolting from the premises.

But all in all, I had a good time during my first year at Chung Chi College. In addition to math, I studied Chinese, English, Japanese, physics, and philosophy. In the latter subject we not only learned about the great philosophers but also learned about what a student (or person, in general) is supposed to be and do. It was a small college, so we all got to know each other. Because we were near the ocean, we often swam or played games on the beach. And what's not to like about that?

That first year was fun, but the second year was more exciting, and things became more serious. CUHK was starting to grow, and Chung Chi College with it. The president, Mr. Lee, who came from the University of California, Berkeley, was intent on building up CUHK. As part of that effort, he added a number of new PhDs to the faculty, including Stephen Salaff, a young mathematician who had come to Chung Chi from Berkeley as well.

Salaff was the first professor I ever had who was really up on modern mathematics. He taught our class on ordinary differential equations in the "American style," encouraging students to speak up and participate at all times—an approach the Chinese students, myself included, weren't used to. Instead, we'd been encouraged to quietly absorb knowledge without interrupting the teacher's train of thought. Because of Salaff's freewheeling style, our classes with him were less scripted and more spontaneous, though he sometimes got stuck in the middle of a presentation. I helped him out at these junctures whenever I could, and he soon took notice of me. Sometimes he let me teach a portion of the lesson, if he felt I was up to the task. I also went to his home quite often, helping him prepare his lecture notes for the class or suggesting a different way of tackling a math problem.

At some point, Salaff realized that these lecture notes, when put together, could provide the basis for a book, which we started working on together. It was hard to get the book published because the preface made it clear that I was still just a teenager. But we did publish it many years later after I had become an established mathematician. And I learned a lot in the course of writing that book, especially after having read through the literature so extensively.

Salaff decided that if I really wanted to pursue mathematics, I ought to study abroad. He was upset that my scholarship money from CUHK was so low, about half of what other students typically received, because I hadn't scored well on my college entrance exam, particularly the Chinese literature portion. He made a big fuss, insisting that I was a talented student who should be getting more money. The university was indifferent to his pleas, but that just made him fight harder.

The college's dean of physical education, a woman called Lu, had also come from Berkeley, and she advised Salaff to abandon this fight, since it was likely to make my situation worse. Knowing that my family was poor, she offered me another way to make money: I could teach tai chi to the school's professors, most of whom were foreigners and therefore unfamiliar with this form of martial arts. Frankly, I wasn't great at tai chi, but it was a pleasant enough way of earning money, and I was grateful to Lu for arranging it.

Another nice development during my second year at Chung Chi was that I was able to get together regularly with faculty and students from the other schools in the CUHK system, including United College and New Asia College. United had just hired James Knight, a very good mathematician from the University of Cambridge whom I came to know. I attended his algebra course, which was excellent, and we built up a nice rapport during the year. At the end of the term, Knight gave me the original copy of his PhD dissertation as a gift before he returned to Cambridge to be a lecturer. Unfortunately, he died about ten years later in a motorcycle accident; I was shaken up when I heard the news, even though we'd long been out of touch.

From the various interactions I'd had with mathematicians like Chow, Salaff, and Knight, word had gotten out that I might be a talented student, at least when it came to math. After receiving a request, from a joint math panel representing all three of CUHK's colleges, that I should graduate early, Mr. Lee (who was now the university's vice chancellor) decided to find out just how exceptional I was (or wasn't). The plan was for me to meet with Y. C. Wong, Hong Kong's most distinguished mathematician. Wong was a differential geometer at Hong Kong University, and he was charged with carrying out the evaluation singlehandedly.

It took me more than an hour and a half to reach Hong Kong University by train, ferry, and bus, followed by a pleasant walk up to the mountains. When I reached his office, it soon became apparent that Wong was not going to test me in any formal way, or in any way at all. He just wanted to talk about his research, which, frankly, didn't sound all that fascinating to me. Wong was working on the geometry of "Grassmannians," which basically involve a space of higher-dimensional planes passing through the origin. He was getting hung up at that time on a calculation that didn't strike me as very difficult. When Wong sensed that I could not appreciate the exciting work he was doing, he reached what was for him the obvious conclusion: I could not possibly be a genius.

I'm not going to dispute his contention, but one might take into consideration the qualifications of the judge. I soon learned that Wong was unable to publish many of his papers on this subject, suggesting that the editors of the math journals in question also failed to appreciate the exciting work he was doing.

But the fact is, I don't like the term "genius," and almost never use it, because I don't really understand what it means. I suspect that some people have a romanticized notion of geniuses as individuals who come up with incredible ideas, or amazing mathematical proofs, virtually out of thin air, as if a vision had suddenly appeared before them. Their intellects are so advanced, according to the folklore, that they can produce these feats without breaking a sweat. In the movie *Good Will Hunting*, for example, the lead character takes a few minutes away from his janitorial chores at MIT to knock off major problems in mathematics. While things like that might happen, I have never seen it. In my experience, solving hard math problems takes hard work, and there's no way around it, unless the problem is rather trivial. If, on the other hand, you work really hard for a really long time and eventually succeed in doing something that no one has done before—and perhaps something that no one thought *could* be done—does that make you a genius? Or just an overachieving drudge? I don't know, but I also think it's not worth spending much time on such questions.

The bottom line is that the powers that be at CUHK concluded I was not a genius—and they got no argument from me. I never contested that verdict. But then again, I've never let that judgment hold me back either.

Nevertheless, my encounter with Wong did nothing to assuage Salaff, who was hell-bent on my graduating early so that I could continue my education abroad and begin what he hoped would be a brilliant career.

I had completed four years of coursework in three years, but CUHK still required four years of study. Vice Chancellor Lee was unmoved by Salaff's entreaties and was unwilling to waive the traditional four-year stipulation. Salaff, however, didn't stop there; he wrote a letter to the newspaper and an article for the *Far East Economic Review,* criticizing the university's bureaucratic approach to this issue. He urged CUHK to show greater consideration for its most gifted students.

Salaff's perseverance was not applauded; several people told him it would be wise to back off. Lee countered, meanwhile, by saying I did not need a degree from CUHK because the famed mathematician Loo-Keng Hua had never gotten a college degree either. I could carry on without a college degree, he said, just as Hua had done.

That piqued both my curiosity and Salaff's on the subject of Hua's educational background. When I read up on him, I found out that he didn't even have a high school diploma. Hua grew up poor in what was then the small town of Jintan, due west of Shanghai. He helped his father, who ran a general store apparently without much success. Hua did math problems on the side, working without any supervision, whenever he had time. He later attended a vocational college in Shanghai, where he won a national abacus competition, but he dropped out of school and resumed working in his father's store when he could no longer afford living expenses at school. Shortly thereafter, Hua published a brief note in a Shanghai scientific journal, pointing out a mistake in a prior article of that journal, which claimed to present a general solution to the "quintic," a fifth-degree equation. Hua's note caught the attention of a mathematician at Tsinghua University in Beijing, which invited Hua to join the department. Hua accepted, starting as a library clerk and eventually working his way up to the status of lecturer. A few years later, he was invited to the University of Cambridge, where he worked under the direction of the famous number theorist G. H. Hardy. Hardy assured Hua that he could earn a PhD in two years, but Hua did not enter a degree-granting program because he believed the registration fees alone would be too expensive for him. He returned to China after two productive years at Cambridge. Although he still did not have a PhD, a college degree, or even a high school degree, his reputation had been established and his career launched.

Salaff was inspired to write an essay about Hua, which was eventually published. Much of the source information about him was in Chinese, of course, which Salaff asked me to translate. So I ended up reading more about Hua, and the more I read, the more impressed I became.

The moral of the story, from CUHK's perspective, was that Hua was able to achieve all that he did in mathematics without the benefit of a college degree—or any other degree, for that matter. The university felt, accordingly, that it did not need to waive its standard rules in order to give me a degree, regardless of how vociferously Salaff objected. For if I was as talented as Salaff claimed, I should be able to overcome a minor impediment like that.

Nevertheless, Chung Chi College did give me a diploma (though not a degree) at the graduation ceremony in June 1969, and practically the whole student body cheered when I received it. It was a small school, as I said, and most people knew about the debate that had ensued regarding my early graduation.

Once Salaff accepted the fact that CUHK would not budge on the college degree issue, he turned his attention instead toward getting me into a PhD program at Berkeley. I asked him whether I should consider other schools, but he thought I should apply just to Berkeley, where he had strong connections to the math department, which was rated among the best in the world.

I saw no reason to disagree. So I took the Graduate Record Examinations (GRE), Test of English as a Foreign Language (TOEFL), and other exams and, fortunately, did well enough. In the meantime, Salaff had written to the mathematician Donald Sarason, a friend of his at Berkeley, touting my promise in mathematics. Sarason sent an application to Salaff, saying that he could probably get me into a graduate program there even though I had not received an undergraduate degree. That gave me some hope. I applied, of course (how could I not?), and heard on April 1, 1969, that I had been admitted. This was among the most important news I had received in my life, and I was elated.

Not only had I gotten into Berkeley, however; Sarason had also lined up the cushiest fellowship available. Funded by IBM, it provided $3,000 per year—a sum that would be very helpful in view of my family's financial straits. I was truly lucky because this situation may have been unprecedented: No third-year student from Hong Kong, as far as I know, had ever started graduate school at Berkeley with such a generous financial package. I'm pretty sure that the mathematician Shoshichi Kobayashi, who was then head of graduate student admissions, and the famous Chinese geometer S. S. Chern, who was also at Berkeley, played roles in securing this fellowship for me. I'm grateful to all four of them—Salaff, Sarason, Kobayashi, and Chern—but most of all to Salaff. Without him I probably wouldn't have made it to Berkeley and perhaps might not have had the opportunity, nor acquired the wherewithal, to leave Hong Kong in the first place.

Chern came to Hong Kong in July 1969 to receive an honorary degree. I had arranged to talk with him during that visit at Hong Kong University, where he'd been invited to give a lecture. I'd read an article about Chern when I was in high school that had called him the most famous mathematician from China—a scholar who was known and respected throughout the world. That was the first time I realized that someone from China could actually be a mathematician of international renown. I hadn't recognized that before because China had a big inferiority complex that lasted a long time, although it was alleviated to some extent in 1957 when two Chinese-born physicists, Chen Ning (C. N.) Yang and Tsung-Dao (T. D.) Lee, won the Nobel Prize in Physics. Yang and Lee's award, coupled with Chern's growing reputation in mathematics, showed that Chinese people could accomplish something on the world stage. Their success and accompanying distinction gave hope to the whole nation—or at least to those of an academic bent.

Yang, in fact, delivered a lecture in Hong Kong in 1964 when I was a high school student. Although I wasn't able to attend the event, I still felt Yang's influence by reading newspaper accounts of his talk. And I now feel lucky to have grown up in an era when prospects for Chinese students were looking brighter than they had just a decade earlier.

It's fair to say that my own prospects were looking brighter too, now that I was headed to Berkeley. Chern knew that his school had accepted me and asked during our meeting whether I was going to go to Berkeley. "Yes," I replied, "I am." That was pretty much the sum total of our conversation at the time, but we would have more to say in the future, for this terse exchange was the start of a long and fruitful—and sometimes difficult—relationship.

Although I was committed to going to Berkeley, just as I'd told Chern, I still had to overcome the usual problem: I didn't have enough money to get there. Nor did I have a visa or an identity card, and getting a visa for the United States was no easy matter. A travel agent with TWA walked me and other students through the process of how to get a visa with the expectation that I would purchase my plane ticket through him. He was angry when I bought my ticket from Pan Am, which offered a better deal.

My mother was happy about my going to Berkeley, while perhaps at the same time a bit worried about my venturing so far overseas. I felt bad too about leaving her alone to take care of my older brother, Shing-Yuk, who'd gotten sick the year before and was subsequently diagnosed with a brain tumor. He'd already had surgery to relieve the pressure in his head. This was yet another trying time for our family, and it made the parting all the more difficult for me. But I also felt a strong desire to make my way in the world, knowing that a number of elements had lined up just right, including Salaff's coming to CUHK, all of which resulted in an opportunity for me at Berkeley. An offer like this might not come again, and I needed to jump on it. I assured my mother that I would remain close to the family in spirit, even though I would be nearly seven thousand miles away, and I promised to write faithfully, sending money every month.

I was excited about my future, having a sense that many doors would open up to me once I landed on U.S. shores. I was a bit apprehensive, to be sure, as I'd never been outside of Hong Kong before, except for the months I spent in China as an infant. In many ways, and on many different levels, this was going to be an adventure. But I was up for it, ready to take on new and bigger challenges at twenty years of age.

I set off for San Francisco International Airport in early September 1969, eager to explore the New World using mathematics as my point of entry, guide, and beacon in my evolving search for truth and beauty. I was traveling light, with just one suitcase and less than $100 in my pocket. I'd left all my friends and relatives behind, as well as all the math books I'd collected over the years, not realizing that my personal library would affect the fate of my younger brother, Shing-Toung, who now goes by the name of Stephen Yau. I was about to start graduate school, while he was just starting college at Chung Chi, where his two older brothers had gone before him.

At American colleges, students often have the luxury of taking a couple of years before deciding what to study. But that's not how it worked in China, or in Hong Kong. Stephen, who was still a teenager, had to decide from the outset what area he'd focus on. My mother weighed in on the subject. "Since your brother left all these math books on our shelves," she said, "the best thing would be for you to study math." So that's what he did, and that shows how things are done in China. Much depends on chance, rather than on a conscious act of will. Fortunately, things have worked out all right, as my brother has done pretty well in mathematics and seems to like it. He's never expressed any regrets to me about his choice of profession—a choice, I suppose, that was only partly his.

Nevertheless, there are worse things one can do, in my opinion, than learning about mathematics and eventually making a career of it. And my brother is, in fact, returning the "favor" bestowed upon him decades ago by purchasing math books in America and donating them to Chinese libraries.

CHAPTER THREE

Coming to America

FROM THE MOMENT I set foot in Hong Kong's Kai Tak Airport on the first of September 1969, everything was new to me. Until that moment I had led a provincial, almost travel-free existence (although I would soon make up for the latter deficit). While I had been to an airport to say goodbye to a friend or relative, I had never been to one for the purpose of my own personal transit. And, of course, I'd never been on an airplane before, not even a fake one at an amusement park or shopping center. Unlike the more jaded travelers on my flight to Hawaii—an intermediary stop en route to San Francisco—I paid close attention to the flight attendants (who were called "stewardesses" back then) as they discussed the safety features of our aircraft and reviewed the procedures to be followed in the event of an emergency.

Fortunately, those procedures were never called for. Stepping off the plane at the San Francisco International Airport, I was dead tired after the twenty-four-hour journey. But I revived instantly when I looked up at a brighter, bluer sky than I'd ever seen before and breathed in air that was cooler and drier than the warm, moist air of Hong Kong's more tropical clime. The weather in California was almost predictably, stereotypically wonderful. For a moment, I wondered whether this was what it's like to arrive in heaven for the first time, not that I'm normally inclined to such reveries. Taking it all in, I reveled in the fact that everything was new to me—the unfamiliar though pleasing vistas, the sky overhead, the ground that I stood on, and even the air I drew into my lungs as I sampled its restorative properties.

I passed through Immigration Control without any trouble, thankful that my Berkeley credentials created a free and clear path forward. Donald Sarason, whom I knew only through written correspondence, met me at the airport. Although I'm not sure what I was expecting, I was initially thrown by his appearance. With his beard and shoulder-length hair, he looked like a hippie—or at least what I imagined a hippie might look like given that there had not been a lot of them in the remote boroughs of Hong Kong I frequented. Not that I'm complaining, because he was an extremely nice, soft-spoken man who went out of his way—and literally drove the extra mile—to make sure I was well looked after on my first night in the United States.

We headed northeast from the airport, crossing the Bay Bridge to reach downtown Berkeley, where we stopped at the YMCA. I needed an economical place to stay, and it was only $10 a night. Once I checked in, Sarason took off, reminding me to drop by the math department the next day—not that I needed any reminders on that score.

In the YMCA lobby, people crowded around a loud TV set, watching baseball—a game I knew nothing about and had never seen before. Nor had I spent much time watching television (although I had caught Neil Armstrong's moon landing a couple of months earlier at a Hong Kong department store). We never had television at home and, in contrast to people who'd grown up in the United States, it had played no part in my life and held little allure for me. I carried my suitcase upstairs to the room I would be sharing with seven others. A large African American man greeted me and asked where I'd come from, apparently surprised by my appearance. I guess I had the look of someone "fresh off the boat," which was not too far off the mark, except for the part about the boat. He slept in the bed right next to me, and I'd never met anyone like him before. It was hard for me to understand his turns of phrase and accent, but I appreciated the fact that the first person I interacted with in the United States, apart from Sarason, was very welcoming.

The Y would suffice for a night or two, or a week, as it turned out. Although the furnishings were bare, it offered me a roof overhead and a bed upon which to sleep—which I desperately needed after my lengthy flight. But it would not do for the long term, as there was nowhere for me to study, so I quickly set about finding an apartment. But the next morning, my first stop was the Berkeley math department. I was warmly greeted by Sandy Elberg, who was then dean of the Graduate Division, and a young mathematics professor named Marc Rieffel. I met another young professor, Tsit Yuen (T. Y.) Lam, who, like me, had come from Hong Kong. Lam was kind enough to loan me some cash, which I repaid as soon as I received the first installment of my fellowship stipend. The short-term loan from Lam was extremely helpful, because I needed to pay the YMCA, secure an apartment, and purchase food, books, and other essentials.

I was advised to check out the accommodations at the International House, located right next to the campus, but no rooms were available. Later, after perusing a bulletin board near the YMCA that had apartment listings, I met three students who were also looking for a place to live. With a four-way split, our monthly rent was $60 apiece. My $3,000 fellowship allotted me $300 per month for ten months, half of which I sent home to my mother. After paying the rent, I had $90 left to cover everything else. There was not a lot of fat in the budget, but I didn't need much.

In the beginning, we each cooked our own food. I had a limited repertoire, tending to eat various permutations of soup, rice, and vegetables. We then decided to share food and try eating dinners together, but that arrangement didn't work out for two reasons: Scheduling was difficult, and my cooking skills were weak, to put it mildly. No one wanted to eat on the nights that I prepared the meals, and that hasn't changed much in the decades since. I may have some strong points, but cooking, music, and physical education are apparently not among them.

I normally woke up around 7 a.m., washed, grabbed a quick bite, and hurried to school. It took me about twenty minutes to walk from my apartment to Campbell Hall, where the math department was then housed. (The next year it moved to Evans Hall.) My usual route took me up the famous Telegraph Avenue, which was often crowded with odd-looking people dressed in colorful, unorthodox garb. Many of them were panhandlers, asking for "spare change," but I ignored their entreaties, as I had very little change and none to spare.

Upon arriving on campus, I shut myself up in the math department's classrooms, library, and lecture halls all day long. If there's any truth to the saying, "All work and no play makes Jack a dull boy," then I was a very dull boy indeed. Furthermore, it was quickly apparent that my training had not been too demanding at tiny Chung Chi College, where I'd been a relatively big fish in a very small pond. Berkeley, by comparison, had an enormous math department with a wide-ranging array of course offerings. Eager to make up for lost time, I went for total immersion—soaking up as much knowledge as I could possibly fit into my normal-sized head. While my exact destination was unclear, I was guided by a saying from the Chinese poet Qu Yuan (circa 340–278 BCE), whom my father liked so much: "The road is long and tedious, but I will follow this road unceasingly until I find the truth." I didn't mind that the road ahead might be long, but I surely hoped it would not be tedious.

If I needed additional inspiration, I could find it in the words of Confucius, who predated Qu Yuan by a couple of centuries and was always good for a rousing quotation: "I have spent a whole day without eating and a whole night without sleeping in order to think, but it was of no use. I got nothing out of it. Thinking cannot compare with studying." I suspect that Confucius would have been proud of me in those days, because I studied so long and hard I barely had time to think.

Although I officially signed up for three courses, I audited six others and attended as many lectures and seminars as I could pack into a day. The resources at Berkeley were staggering; there was a large group of outstanding mathematicians on the faculty, and excellent students as well, including my classmate Bill Thurston, a future winner of the Fields Medal—often regarded as the "Nobel Prize of mathematics." In addition to the regular courses it offered, the department also hosted special lectures and seminars several times a week.

Like a half-starved person showing up for the first time at an all-you-can-eat buffet, I tried to take it all in. I adopted this nonstop approach because I wanted to and because I could: I didn't know many people, had almost no social obligations, and had very few competing demands on my time. Mathematics was my sole focus, at least in this early stage. That's what I had crossed the ocean for, and it consumed most of my waking hours. I started classes at 8 a.m. and kept at it the rest of the day, sometimes having only five minutes to get from one class to another, which might be at the other end of campus.

I often didn't have time for lunch, in which case I ate a sandwich during a lecture, preferably while sitting in the back of the room so as not to distract others. I took classes until about 5 p.m. and then I walked home, stopping at the big university bookstore on the way to browse through the newest math books. Near my apartment was a supermarket, where I often picked up a few items before heading home for the night. Life, you might say, was very simple in those days: It started with math and ended with math, with a fair sampling of math filling out the middle, as well.

During the first semester, I took algebraic topology (taught by Edwin Spanier), differential geometry (taught by Blaine Lawson), and differential equations (taught by Charles Morrey). I also sat in on courses in algebra, number theory, group theory, dynamical systems, automorphic forms, and functional analysis.

The three courses I enrolled in ended up having a big influence on me. Before coming to Berkeley, I thought I knew about topology, which involves the study of shapes in the most general sense—as well as patterns and categories of shapes—but Spanier's course on algebraic topology, which transformed topological problems into algebraic ones, offered a whole new take on the subject. I was nervous in the beginning, because students were expected to interact more in class than I was used to. I wasn't prepared to say much at first, whereas many other students were much less hesitant and seemed to know what they were talking about. A couple of weeks later, after having gotten through a big chunk of the textbook, which was written by Spanier himself, I realized that most of the students were just spouting nonsense—showing off, in other words.

Lawson's course helped spark my interest in geometry, which is like topology in that it also focuses on shapes but in a much more specific way. In geometry, which concerns the exact shape of an object, a sphere and a cube are wholly distinct. In topology, however, a sphere and a cube belong to the same class of objects—and are equivalent, in other words—because one can be deformed into the other with some bending and stretching, but no cutting or tearing.

Back in Hong Kong, I had viewed mathematics as an abstract subject. For the most part, I was on my own then, clinging to the rather unformed opinion that the more abstract areas of the discipline were somehow better—that they were purer and closer to the essence of mathematics

A sphere, cube, pyramid, and tetrahedron are distinct shapes insofar as geometry is concerned. But in topology, these seemingly different shapes are considered equivalent because each can be fashioned from (or into) the other by bending, stretching, or pushing, without having to resort to any tearing or cutting. (Based on original drawings by Xianfeng [David] Gu and Xiaotian [Tim] Yin.)

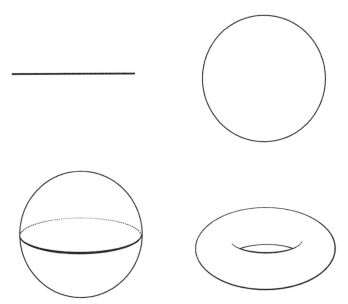

In topology, there are just two kinds of one-dimensional spaces that are fundamentally distinct from each other: a line and a circle. You can make a circle into all kinds of loops, but you can't turn it into a line without cutting it. Two-dimensional compact surfaces that are "orientable"—meaning they have two sides like a beach ball, rather than just one side like a Möbius strip—can be classified by their genus, which can be thought of, in simple terms, as the number of holes. A sphere of genus 0, which has no holes, is therefore fundamentally distinct from a torus, or donut, of genus 1, which has one hole. As with the circle and line, you can't transform a sphere into a donut without cutting a hole through the middle of it. (Based on original drawings by Xianfeng [David] Gu and Xiaotian [Tim] Yin.)

and therefore closer to the "truth" itself. I assumed that I would concentrate on an abstract subject like operator algebra, which was a branch of functional analysis I'd become intrigued with at Chung Chi in consultation with Elmer Brody, a lecturer in mathematics. I'd read many books on functional analysis and had even written to Richard Kadison of the University of Pennsylvania and Irving Segal of MIT, asking for reprints of their papers without knowing they were top authorities on the subject. When I met them many years later, they treated me like an old friend. Segal even took me out to dinner.

They were nice to me, despite the fact that I had not gone into their field. My feelings toward the subject changed soon after I arrived at Berkeley. For one thing, after sitting in on some seminars on functional analysis that fall, I just wasn't as excited about it as I had been. I found the other courses I was taking more interesting, which prompted me to back away from my bias toward abstraction as the most important criterion.

Instead, I came to regard mathematics less as a self-contained discipline, or subject unto itself, and more as a field of study that is closely aligned with nature. I was able to see the connections with nature more readily through geometry and the beautiful structures that spring from it. In some cases, we can even draw pictures of these structures, which makes it feel more palpable, though that is hard, if not impossible, to do in the more esoteric realms of mathematics.

Over time, I found myself drawn to geometry more and more for that very reason. The subject was deeper and richer than I had realized before, based on my earlier, superficial encounters. And this venerable field—which dates back twenty-five hundred years to the time of Pythagoras and four thousand years to the time of the ancient Egyptians and Babylonians—pulled me right in.

That said, Morrey's class on differential equations probably had the biggest impact on me of all. The focus was on *partial* differential equations—equations in which something can vary with respect to multiple variables, not just with respect to a single variable such as time. These equations are incredibly important because, among other reasons, the major laws in physics—those formulated by Newton, Maxwell, and Einstein—are written in the form of partial differential equations. The "nonlinear" form of these equations are particularly challenging. Most cannot be solved exactly or "explicitly." Solutions can be arrived at only through an arduous process of approximation.

The course relied heavily on Morrey's textbook. In some ways, it was poorly written, as the material was not well organized. Yet the content was very good, and I still think it's the best book available on the subject, despite its shortcomings. The course, however, was not popular because word had gotten out that the material was extremely difficult. Morrey also asked students to present material in front of the class, which could make for an uncomfortable experience—both for the presenters and presentees. I stuck with the class because I knew in my heart it would be important for me. I worked hard, doing a ton of calculations, and learned a lot in the process.

Somewhere in the back of my head, I had vague notions about tying together geometry and topology, using partial differential equations as the connective thread. Geometry and topology are often viewed as separate subjects, but the divisions between the two always struck me as artificial. Geometry can give you a close-up view like the one you get from holding a magnifying glass to the Earth's surface, whereas topology gives you the kind of big-picture looks you can get only from outer space. But in the end, it's the same planet, and the two views should thus be considered complementary rather than competitive.

So I never understood why anyone would try to draw a line between geometry and topology and isolate the two disciplines, as some practitioners seemed keen on doing. There's no need to pick one over the other when they should go hand in hand. In fact, I see all of mathematics as part of the same fabric and have never felt constrained by the boundaries artificially imposed between subjects. I'm interested in all of it—"the whole enchilada," as my American friends sometimes say—propelled by an overriding sense that, as we come to understand the different parts better, we find they do indeed hang together. Yet I still acknowledge that some of those parts, for mysterious reasons, simply appeal to me more than others.

To be clear, I was by no means the first person to have had thoughts along these lines. The Gauss-Bonnet theorem—which was developed over much of the nineteenth century through the collective efforts of Carl Friedrich Gauss, Pierre Bonnet, and Walther von Dyck—related the geometry (or curvature) of a particular type of surface to its topology. Early in the twentieth century, Henri Poincaré firmed up the link between geometry and topology, and some decades later Heinz Hopf and S. S. Chern (who became my mentor at Berkeley) made that link even stronger. I was merely trying to build on what they had done by bringing differential equations, particularly of the nonlinear partial variety, to bear in this endeavor. My early forays along these lines became part of a field that was eventually called "geometric analysis"—a term coined by the American Mathematical Society and National Science Foundation in order to categorize research projects in this area.

The new element here was the attempt to make use of nonlinear partial differential equations, because differential geometry, which applies the tools of calculus to geometric problems, had been around for a couple of centuries, dating back at least to the work of Leonhard Euler in the mid- to late 1700s. The basic premise of that venerable field, bringing at first linear differential equations into geometry, makes perfect sense because these equations describe how things change on tiny, infinitesimally small scales. In geometry, we use such equations to measure curvature and to figure out how curvature changes as we move about a space. By determining the curvature of a space "locally"—in one small portion, that is—we can learn about the entire, "global" space. That link—between curvature, the local geometry or precise shape of a given space, and topology, the general shape of that same space—has long fascinated me, lying at or near the center of my research over the past forty-plus years.

At their very essence, both geometry and topology are concerned with shape, and curvature provides a means of determining that shape. A fully inflated soccer ball, which assumes the form of a sphere, is topologically equivalent to an underinflated, bashed-in soccer ball of the same size. In this case, one shape (a perfectly round ball) can be transformed into the other (a dented ball) simply by adding or withdrawing air; no tearing or cutting is necessary. Whereas the spherical ball has constant (positive) curvature, which does not change from point to point, the curvature of the deformed soccer ball varies at different points on the surface.

Curvature, again, is the key to getting a fix on both the overall shape (or topology) and the exact shape (or geometry), and this connection also holds for higher-dimensional objects, characterized by different kinds of curvature, that are far more complicated and much more difficult to picture (or kick) than soccer balls with varying degrees of air pressure. That's part of the reason why curvature is such a powerful gauge, and one that has held my interest for so long.

While we could define a two-dimensional sphere, for example, simply as the set of points in three-dimensional space that lie a certain distance from a central point, we could also define the same object solely in terms of its curvature properties. The latter approach, it turns out, is more powerful than the former, having much broader utility: It can be used to describe more complicated and convoluted objects (or manifolds) in higher-dimensional spaces—cases for which simple formulas are not available.

Curvature also plays a big role in physics, which is built upon laws that are governed by differential equations. A particle's velocity, for example, depends on how its position changes with time. A particle's acceleration depends on how its velocity changes with time and so forth. We can, for instance, figure out the force on a moving particle, and hence its acceleration, by determining the curvature of its trajectory. In high-energy accelerator experiments, researchers can work in the opposite direction too—figuring out a particle's mass, and hence its identity, by analyzing the curvature of its path. And those are just some of the many ways in which curvature comes into play in physics. (On a metaphorical level, one might think of an individual's personal trajectory and, judging from the "curvature" of that path at various key junctures, learn something about the shape of a person's life, which is roughly what I'm trying to do in this humble account.)

Looking on a far broader scale, the Einstein equations from the theory of general relativity, which I would learn about later that year, describe the curvature of the universe itself. They consist of a set of differential equations of the "nonlinear" variety—equations in which small changes in one variable can have disproportionately large consequences. Many phenomena can be approximated reasonably well by linear equations—"linear" meaning not only that changes are proportional, but that if you have two solutions to the same equation and add them up, their sum is still a solution. Nevertheless, the world we inhabit is intrinsically nonlinear, and that fact cannot be forever ignored.

Nonlinear equations are needed, therefore, to understand sudden shifts in climate or wild fluctuations in the stock market. These equations populate the realm of general relativity too, where space is always curved and associated phenomena are inexorably nonlinear. One idea I would soon be grappling with, in keeping with a general strategy in geometry, would be to use the equations of general relativity, which describe things on a local scale, to try to understand the global structure of the universe.

The catch, of course, is that nonlinear equations are notoriously difficult to work with. But I had landed, by chance, in the classroom of Morrey, probably the world's leading expert in the field of "nonlinear analysis"—analysis being an advanced form of differential calculus—and his specialty was nonlinear partial differential equations. I was eager to take in as much as Morrey was willing to dish out. Fortunately, he was generous in this regard.

I was buoyed in these efforts by a growing realization that combining geometry, topology, and nonlinear analysis in just the right way could be extremely fruitful. At that time, the work of people like Morrey in partial differential equations was largely separated from the work of people in geometry, even people in his own department like Chern. Many geometers, including Chern, were content to leave the partial differential equations to the analysts—or, as one noted authority put it, to the "engineers." While Morrey was a first-rate analyst, he was not especially interested in geometry per se. Instead, he saw geometry more as a source of interesting partial differential equations for him to play around with. I, on the other hand, was hoping to flip that around, using those same equations to solve problems in geometry—problems we hadn't been able to make progress on by other means.

I sensed that bringing these separate strands together could yield huge payoffs for geometry and analysis, as well as for topology. My ideas were half-baked at first, as I didn't really know how to proceed or where to go. But my thinking gradually firmed up, and I've held to my conviction ever since.

But we're running a bit ahead of ourselves. In the fall of 1969, the Vietnam War protests were in full swing, and Berkeley was a focal point for student dissent. Many students and faculty went on strike. Spanier canceled his class when not enough students showed up to carry on. The students in Morrey's differential equations course went further than missing a few classes; everyone dropped out except for me, who was brand new to the country and not yet enmeshed in the political debate. Yet Morrey was still willing to continue the course. He wore his usual coat and tie, and he kept delivering lectures to me, just as he would have to the entire class. In fact, he did more than just the usual preparation. Rather than designing the standard course lecture, he crafted lectures that were tailored specifically for me, geared both to my interests and to my level. I never expected one-on-one attention like this at a big university like Berkeley, with thirty thousand or so students, but these truly were extraordinary times. And I found myself in a fortunate position, learning the tools of the trade from the master himself.

Berkeley was the site of large, frequent, and boisterous protests. The smell of tear gas hung in the air so often it was almost part of the ambience. It was not unusual for me to look outside during a class to see large crowds of students, with rocks in their hands, facing off against police holding shields and guns. "The whole world is watching!" antiwar demonstrators sometimes chanted. And I was watching too—not on a TV screen, but rather from inside the classroom or library. It was, admittedly, hard to focus on math with all the mayhem unfolding on the other side of the glass. But I wasn't ready to get personally involved in this struggle, even though I was no fan of war, mainly because I was not yet a part of the American culture and hadn't had a chance to process all the issues at play here.

Salaff returned to Berkeley that year, after a brief stint in Japan, and tried to introduce me to American ways. He took me on some sightseeing trips in and around San Francisco. He also invited me to some gatherings at his house where marijuana was passed around freely. People were generous about sharing and always asked me whether I wanted a "hit" or a "toke." But I always declined, never once trying marijuana, even though it was ubiquitous throughout Berkeley in those days. From Salaff and his friends, I got some sense of how hippies behaved. The free-spirited lifestyle they exhibited bore little resemblance to what I'd seen or experienced growing up in the more straight-laced environment of rural Hong Kong. Most people I encountered there were hard-pressed to scrape together the necessities of life; recreational drugs rarely entered the equation.

That said, I didn't judge anyone and was friends with many people who might be labeled hippies. But I never sported that particular look myself and never got into the drug scene, either. Nor did I get anywhere with alcohol, although Salaff, as an older and more experienced friend, thought it might be time for me to learn how to drink. My first chance came at a math department picnic, which was held in Tilden Park, high in the Berkeley Hills. Morrey had specifically encouraged me to come. They were serving beer, so I picked up a tall glass and downed the whole thing very quickly. Within about ten minutes, I got extremely dizzy and told Morrey I'd better go home. He volunteered to drive me. I got back to my apartment around 3 p.m. and went to sleep, not waking up until noon the next day. That's when I learned how sensitive I am to alcohol. I've been cautious around it ever since, only drinking small quantities, if anything.

I had met some Americans in Hong Kong through one of the churches my mother went to for relief during lean times—a period that could in fact apply to most of my childhood. Part of this family lived in Berkeley, and they invited me to their house for Thanksgiving. I had no idea what Thanksgiving was but knew it must be a big deal, given how abandoned the campus became in late November. A lot of people showed up at their house for the holiday dinner; some were clearly family members while others appeared to be random stragglers like myself.

I was told beforehand to bring an item that cost no more than $1 for a gift exchange. I bought some kind of crystal object I found at a variety store, putting it on a table with the other gifts. No one wanted the thing I got, which was a little embarrassing, though I had no use for it either. I didn't find out much about Thanksgiving that night, but I did eat a lot of food. And maybe that was all I needed to learn.

Christmas came around soon after that, which I'd never celebrated before. I didn't celebrate it this time either, but I did find out how serious Americans are about this holiday, too, because, once again, the entire campus cleared out. I was about the only person there for two weeks. Fortunately, the math library stayed open the whole time (except for Christmas Day), which was my Christmas Miracle. And I pretty much moved in.

I'd already passed many hours in the math library because first-year graduate students don't have offices, so it effectively became my office. I spent most of my free time there when I wasn't in class. In those days, very few journals were devoted to mathematics compared with the number available today (presently estimated at around two thousand). So I used to pick up every math journal in the library's collection and try to read the papers. Even if I couldn't understand them fully, at least I would know who wrote what. That gave me a broad sense of what was going on in the field, which enabled me to draw a general picture in my head of how the various strands of activity might fit together.

During that Christmas break, I virtually had the library to myself, with one notable exception: On one occasion, I saw a beautiful young woman about my age who struck me as being almost certainly Chinese. She had come to borrow some books, and I was instantly taken by her. I tried not to stare too obviously, but it was hard to be inconspicuous when hardly anyone else was in the room. Despite my keen interest, I never approached her, nor did I say a word, for that is not the Chinese way. One must wait for a formal introduction.

Some time after the term resumed, I learned she was a physics graduate student, living in the International House nearby. Other than that, I didn't make much headway. I'd occasionally see her at the math colloquium, which was always held in LeConte Hall in the physics building next door. But, again, we never spoke. I had to refrain until the right, and proper, opportunity arose, which took about a year and a half, but it was well worth the wait. For that marked the beginning of a long, off-and-on courtship that ultimately culminated in our marriage.

Apart from that one moment in the library when I first laid eyes on my future wife, things were a lot slower. It was just me and the books. The library, in fact, had a whole shelf of books by Leonhard Euler, the great Swiss mathematician from the 1700s, which I would have read were they not written in Latin—a language that was, you might say, Greek to me. But I continued to read through the journal articles.

I came across a recent paper by the Princeton mathematician John Milnor titled "A Note on Curvature and the Fundamental Group." But this time I did more than just read his paper. I started to think that maybe I could expand on some of Milnor's ideas. Some of the impetus for this thought might have been that I was alone in the library with plenty of time on my hands and little else to do. But the paper impressed me, sparking something in me I had not felt before—the sense that I might be able to do some original mathematics.

I looked up a theorem proved by Alexandre Preissman that Milnor

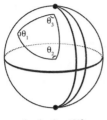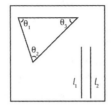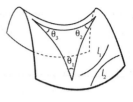

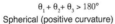

$\theta_1 + \theta_2 + \theta_3 > 180°$ ..... $\theta_1 + \theta_2 + \theta_3 = 180°$ ..... $\theta_1 + \theta_2 + \theta_3 < 180°$
Spherical (positive curvature) ..... Euclidean (zero curvature) ..... Hyperbolic (negative curvature)

On a surface with positive curvature such as a sphere, the sum of the angles of a triangle is greater than 180 degrees, and lines that appear to be parallel (such as longitudinal lines) can intersect (at the North and South poles of a globe, for instance). On a flat planar surface (of zero curvature), which is the principal setting of Euclidean geometry, the sum of the angles of a triangle equals 180 degrees, and parallel lines never intersect. On a surface with negative curvature such as a saddle, the sum of the angles of a triangle is less than 180 degrees, and seemingly parallel lines diverge. (Based on original drawings by Xianfeng [David] Gu and Xiaotian [Tim] Yin.)

had mentioned in his paper. Preissman's theorem pertained to spaces that have "negative" curvature, such as the top of a saddle you might strap to a horse. If you construct a triangle on a saddle, or another object with negative curvature, by connecting three points with the shortest possible paths between them, the angles of this triangle always add up to less than 180 degrees. (On a space of zero curvature, such as a flat sheet of paper, the angles of a triangle add up to exactly 180 degrees, whereas on a sphere, which has positive curvature, the sum of the angles is greater than 180 degrees.)

Preissman considered two closed loops sitting in a space of negative curvature. You start at one point and follow a path that eventually brings you back to the original starting point, which we can call loop A. From there, one can follow another winding path that returns you to the same point, which we'll call loop B. Preissman showed that in such a space, a composite loop made by going around first A and then B cannot be deformed to the composite loop made by going around B and then A— unless A and B follow the same closed path. That one exception is what we call the "trivial" case.

I extended Preissman's theorem to a more general situation, spaces of "nonpositive" curvature, which includes spaces of negative and zero curvature. To prove the nonpositive curvature case, I had to bring in group theory. The definition of a group, in this context, is fairly simple—a set of elements, containing both an identity element (such as 1) and an inverse element (such as $1/x$ for every x), to which certain operations can be performed (such as multiplication) and various rules apply.

In this case, I had to deal with groups that contain an infinite number of elements, about which not much was known then (or even today), although I had, once again, learned from an important paper Milnor had written on that subject. I also recalled a conversation I had during teatime at Chung Chi College with Professor Ronald Francis Turner-Smith. I asked him what he had worked on at the University of London, and he mentioned something about groups of infinite order. I can't remember much of what Turner-Smith said, but he did mention an earlier paper by Issai Schur and Richard Brauer, which I felt might bear on my current problem. I spent all day in the library, searching through old mathematical journals, until I found a paper by Schur and Brauer that was just what I needed. I wasn't interested in group theory when Turner-Smith told me about the paper, but I never would have looked for it had we not talked, and that paper really saved me.

The moral of the story, I suppose, is that casual conversations may end up being more important than you realize. And sometimes you have to remember only a sentence or two of what someone says, be it at a lecture or colloquium or teatime. In this case, a chance remark, which somehow lodged in my brain, enabled me to complete my first proof of any significance.

Although the result I obtained was not earth shattering, the thing I liked most about my proof was the same thing I liked about Preissman's proof: Both showed how the topology of a space (its general shape, in other words) could affect, and constrain, the geometry (or exact shape) of that same space. This is an avenue I have continued to pursue, and it has been a productive avenue for me, as well as for others in geometry and topology.

I went over my proof as many times as I could stand to look at it, checking and rechecking every step, and the argument appeared solid. When school vacation was over, I showed what I had done to Lawson, who was teaching my geometry course. He agreed that my work looked good, and together we went on to prove something else that was loosely related to my theorem and Preissman's. Lawson and I showed how topology could answer the question of whether a space of nonpositive curvature could be the "product," or combination, of two different spaces.

Lawson was eager to submit both papers, and we sent them to the *Annals of Mathematics*—considered by some to be the top U.S. journal in mathematics. Since I did my first proof during Christmas break, completely cut off from everyone else, I did not realize that the thing I proved on my own was actually a conjecture that had been originally posed by Joe Wolf, a Berkeley mathematician and former Chern student who was then on leave. I knew of Wolf already, even though I had not yet met him, because I had read his book, *Spaces of Constant Curvature,* and admired it.

A further coincidence was that the thing Lawson and I proved had been proved independently by Wolf and his colleague Detlef Gromoll, although their paper had not been published yet either. When we met with Wolf, he was not thrilled by the fact that we had done such similar work. Lawson and I were also dismayed to find out we weren't the only ones to have proved this rather obscure point. On the other hand, when we embarked on our project, we had no way of knowing about Wolf's work with Gromoll.

Chern, however, was relieved to hear that someone he'd helped bring to Berkeley (namely me) had done something of note during his first semester of graduate school. Maybe the department's investment in me would pay off. And I was happy too, having contributed something new to mathematics, if only in a minor way.

The *Annals* accepted my solo paper but rejected my joint paper with Lawson. He was upset because, having just earned his PhD two years earlier, he felt it was difficult for new PhDs to break into the field's top journals in the face of competition with more established mathematicians. The good news is that Lawson and I then submitted our paper to the *Journal of Differential Geometry,* and it was accepted. I believe that Chern might have put in a favorable word for our joint effort, which surely would have helped its chances.

The year 1970 turned out to be a memorable one for me. I'd gotten my first taste of the mathematics publishing world—and of the elation one can experience when a paper is accepted, the disappointment felt when a paper is rejected, and the tension that can sometimes emerge over questions of priority and credit.

The spring quarter of that year was anything but tranquil: There were even more student uprisings against the war when news of the U.S. secret bombing campaign in Cambodia came out. Classes at Berkeley were once again interrupted by schoolwide boycotts. Lawson held a geometry class at his home to avoid overtly violating the boycott. These classes lasted only a few weeks. His wife may have objected to having their house taken over by a group of cantankerous students, revved up by what was being called "the war at home." And to her it may have seemed at times as if that war were taking place inside her home.

I continued to work with Lawson throughout the winter and spring. He was a lecturer then and had to share an office in rather crowded, temporary quarters, which didn't offer a great place for us to meet. Instead, we talked frequently on the telephone when he was at home. These conversations could be lengthy, lasting an hour or two and sometimes more. Lawson got divorced a couple of years later, and I feared that these phone calls contributed to the rift. But his former wife later assured me that other issues were more significant, and the divorce had nothing to do with me.

Around the same time in 1970, I sat in on some talks on general relativity given by Arthur Fischer, who was then a lecturer in mathematics. I'd already had one encounter with Fischer: I was making a photocopy of a draft of my *Annals* paper, and he asked to see it. I didn't hand it over immediately, as I was shy about showing my work to strangers, especially someone who looked to me like a wild hippie. Fischer grabbed the paper from me and quickly scanned it, declaring that "anything that relates geometry to topology should be important for physics." I'd already been impressed from Milnor's work about the value of relating geometry, or curvature, to topology, but I didn't know much about physics then or how it might tie in. When Fischer affirmed, without hesitation, that the link between curvature and topology is relevant for physics too, I got excited, because I was already becoming fixated on that same link. I wanted Fischer's statement to be true, but it took many years—until after my proof of something called the "positive mass conjecture"—before I knew that he was right.

That "wild hippie," it turns out, had a surprisingly big impact on me, although I showed up at his first lecture simply out of curiosity, with little in the way of expectations. I had never studied general relativity before—a theory that encapsulates our current understanding of gravity, as dreamed up by Albert Einstein more than a century ago. Einstein's theory, in turn, was built upon geometric techniques that had been developed sixty years earlier by Bernhard Riemann. I figured it would be worth learning something about this subject, as I'd heard the term "general relativity" on countless occasions without knowing much about what it meant. I had no idea how important this subject would soon become for me and for my career.

Gravity, according to Einstein, was not really an attractive force between two or more massive objects (as Newton's law held) but rather the distortion, or curving, of space due to the presence of massive objects and other effects. This picture can explain the motion of planets around the sun, as well as more subtle effects that the conventional, Newtonian view of gravity could not account for. To paraphrase the Princeton physicist John Wheeler: Mass tells space how to curve, and space tells mass how to move. A key term in Einstein's equations, the Ricci curvature tensor, determines how the distribution of matter in the universe affects the curvature of space.

In the middle of one of Fischer's lectures, all kinds of ideas started to flood my head. By then, I was getting more and more interested in geometry, which has a lot to say about curvature, including many different kinds of curvature that are not easily discernible (or discernible at all) in everyday experience. I wondered: If gravity is the result of mass telling space how to curve, as physicists sometimes describe things, what happens in a space that is wholly devoid of matter—the kind of space we call a vacuum? In other words, can a space without matter still have nonzero curvature and gravity?

I kept coming back to this question without realizing that the geometer Eugenio Calabi had asked an almost identical question in 1954, couching his "conjecture" in complicated, mathematical language I won't attempt to define here—complex, Ricci-flat manifolds endowed with a vanishing first Chern class and Kähler geometry—terminology that seemed to have nothing to do with gravity at all. Calabi acknowledges that he was not thinking about physics when he first posed his conjecture. The conjecture applies to spaces with a special kind of geometry, Kähler, which in turn implies a special kind of symmetry sometimes called "supersymmetry." Putting it into nontechnical terms, Calabi wondered what the lengths of various paths within a Kähler space, which can help characterize the space, have to do with the density of that space. The density, in turn, relates to a property called the "volume element," which can be used to determine the space's volume. Calabi also wondered, conversely, what the volume element (or density) of a Kähler space had to do with path lengths—and the notion of distance—within that same space.

You can imagine, for example, learning about a sphere by measuring the distance between a set of points on its surface. But how do you go about measuring distance and volume in higher-dimensional spaces—spaces of, say, six or more dimensions?

Calabi's focus, on mathematics and nothing but mathematics, was not unusual at the time he concocted his conjecture. Even in 1970, as I sat through a lecture on physics presented by the mathematician Fischer, math was pretty far removed from physics. Many mathematicians viewed their subject as "pure" and shied away from anything that smacked of being "applied," including physics.

Distinctions of this sort have not always been drawn throughout history. The scientists of ancient Greece, for example, did not regard mathematics and physics as separate disciplines. And many great mathematicians over the years—including Euler, Gauss, and Poincaré—had no qualms about working in astronomy and other disciplines. Although I was new to the field, and had yet to do anything of real consequence, I still sensed that work in mathematics—especially within the areas that interested me—had the potential to connect with physics in profound ways, even though my grasp of physics was still quite thin. I felt that these ideas would lead somewhere and hoped it would be somewhere interesting.

Over the years I have often skirted the line between math and physics, finding that an exciting and profitable place to be. Mathematics, however, has always been my home base, primarily because it strikes me as the deeper, more fundamental of the two disciplines, for this reason: Any theory in physics needs to be validated by experiment, and results in physics are always subject to change in light of new empirical evidence. On the other hand, when a theorem is proved in mathematics—assuming the calculations are sound and the logic airtight—that statement will always be true. Eternal truths are hard to come by in science, and indeed in any sphere of life, and I believe that has a lot to do with why I gravitated toward mathematics.

But back in 1954, when Calabi published his conjecture, I was just five years old—a hungry boy growing up in Hong Kong. Sixteen years later, as I sat in a Berkeley lecture hall, I was still hungry, though in a different way. I was hungry to devour mathematics and learn enough so that I might eventually take on one of the big challenges the field had to offer.

During my relentless reading in Berkeley's library, I began to dig up everything I could about Ricci curvature. In the beginning, I had not yet come across the name Eugenio Calabi, nor did I know anything of his work. But I soon found references to him during my review of the Ricci curvature literature, and it didn't take long for me to find the 1954 conference proceedings that included his conjecture.

That paper struck a chord with me. I became convinced that the Calabi conjecture held the key to understanding Ricci curvature and how it relates to geometry. Regardless of whether the conjecture was true or false, I believed that either way, its solution would unlock the mysterious structure of Ricci curvature. I believed, more generally, that if we could not solve this problem, we would be unable to solve a slew of problems in geometry related to curvature.

For in higher-dimensional spaces, different kinds of curvature come into play, and among them, Ricci curvature is perhaps the most enigmatic of all. At that time, very little was known about this type of curvature, despite the important role it played in the theory Einstein had conceived more than a half century before.

I was drawn to the Calabi conjecture because of my interest in Ricci curvature—both for its own sake and because of its relevance to general relativity. And I felt that I might be able to carry the ball further, as I did in my work related to Preissman's theorem, once I figured out the best way to approach the problem. But one thing was apparent from the outset: this would be a long-term project, not the kind of thing that could be knocked off during a school vacation. If I were to have a chance of proving the conjecture, I would have to proceed systematically, patiently laying the groundwork.

In the meantime, there were some pressing matters that I, as a first-year graduate student, simply had to attend to. The first order of business was my PhD qualifying exam, which I took early in 1970. It was an oral exam divided into three parts: geometry and topology, analysis and differential equations, and algebra and number theory. The topology test was given by two professors, Emery Thomas and Alan Weinstein. Thomas started off by asking some fairly easy questions about topology that I had no trouble with. He then asked some tricky questions that got very technical. I should have admitted that I didn't know how to answer some parts but instead blundered ahead.

Weinstein began, like Thomas, with some elementary questions about geometry. That part went smoothly. But then he started fixating on some exceptional cases of various theorems and, again, I didn't handle it well. I got a B+ overall—adequate, perhaps, though nothing to write home about.

The analysis and differential equations segment was administered by Morrey and Haskell Rosenthal. I did better on this part, getting an A. The last exam was on algebra and number theory, two subjects I hadn't done much work on before. Somehow, I impressed the professors—Manuel Blum, Lester Dubins, and Abraham Seidenberg—so much that I got an A+. It is rather ironic that the test scores went exactly the opposite of what I ended up doing in my own research. But the good news is that I managed to pass the qualifying exam and thus had one big hurdle out of the way.

Also around that time, the math department agreed to continue my fellowship for another year, which, as I've said, was the most generous fellowship available in the department. That was a huge relief for me, as I'd been unfailingly sending half of the money home to my mother. I also did not have a green card (permanent resident card), which meant I could not obtain support from the National Science Foundation. As a result, I was extremely dependent on this fellowship and was grateful when it was renewed.

Next on the agenda, I had to start thinking about my dissertation and select a dissertation supervisor. I had kept up my relations with Morrey, and toward the end of the spring quarter he asked me to become his PhD student. While mulling over that offer, which I seriously considered, I also spoke with Chern when he returned from a sabbatical around June 1970. I ultimately decided to become Chern's student when, in a moment of clarity, it dawned on me that I liked geometry more than any other subject in mathematics and therefore ought to work under the tutelage of a world-class geometer.

In the meantime, Morrey's health took a tragic downturn. Less than a year later, he started showing signs of Parkinson's disease, and his condition quickly deteriorated. It was devastating to witness the decline of this great mathematician.

It quickly became clear that by choosing Chern as my advisor I had signed on with a dominant force in the department; he was also widely considered to be the greatest living mathematician of Chinese descent. Although he had made many contributions to mathematics, he was most famous for developing the concept of Chern classes, which provided a convenient way of classifying "manifolds"—topological spaces, like the surface of the Earth, that appear to be flat in the immediate vicinity of every point on that surface. Chern had come to Berkeley in 1960, after having spent eleven years on the University of Chicago faculty. He subsequently strengthened Berkeley's programs in topology and geometry, thereby transforming the department into a world leader.

Chern was not only a great mathematician; he was also very skillful at dealing with people. And he liked to entertain, constantly inviting people to dinner at his home, where his wife happened to be an excellent chef of Chinese cuisine. By becoming Chern's advisee, I had entered this social world.

He had a beautiful house on a hill in El Cerrito, just north of Berkeley, which afforded spectacular views of San Francisco Bay and the Golden Gate Bridge in the distance. Chern even had a gardener to help out with the landscaping, which was quite attractive. I attended a number of dinners and parties there, along with other students and faculty. Two young professors of geometry and topology, both in their early thirties, were regular fixtures at these gatherings—Wu-Yi Hsiang and Hung-Hsi Wu. The algebraist T. Y. Lam also showed up from time to time.

I got a little spoiled when visiting Chern's luxurious abode, but I always had to come back down to reality—to my threadbare quarters in a more urbanized section of Berkeley and the reality that, with summer approaching, my roommates would soon be moving out. As luck would have it, I found a studio apartment on Euclid Street, practically across the street from campus, and the rent was just $90. The other bit of luck was that my friend and classmate from CUHK, S. Y. Cheng, was moving to Berkeley that summer to start graduate school in mathematics, and he, too, needed a place to live. Cheng arrived in June and we shared the studio. It was barely big enough for the two of us, though we made do, and the location was extremely convenient. The only drawback was that we lived right above a bar, which could be very noisy, especially on Friday and Saturday nights, but we were young enough to take that sort of thing in stride.

One way we coped with the noise was to stay up late, often until 4 a.m., talking or reading or working on math. I got on a later schedule than I had adopted earlier in the year, but that was okay because I was no longer taking classes nonstop. Confucius might have been disappointed in me, as I was no longer devoting myself entirely to the study of math. I was also setting aside time for thinking about the subject and wondering where, within the diverse range of possibilities, I might go. That exercise, in turn, led me to something of a new perspective.

Upon reflection, I felt that what I did during the first year with my two published papers was a decent start, but there were limits as to how far I could go applying ideas of geometry to group theory and vice versa. I felt that geometry should have a bigger scope. A more promising direction, I concluded, would be complex geometry, which concerns spaces, or manifolds, that can be described only with complex coordinates—numbers that incorporate both a real component and an "imaginary" component (consisting of multiples of i, the square root of -1). I started attending a seminar on that subject led by Shoshichi Kobayashi, who encouraged me to read a book by the German mathematician Friedrich Hirzebruch, *Topological Methods in Algebraic Geometry*. This turned out to be a very important book for me. I read it by myself and started to absorb the whole subject. Interestingly, I learned about Chern classes from Hirzebruch's book rather than from Chern himself, even though he was my advisor.

The more I read Hirzebruch's book and related papers, the more I realized that this subject had many layers that would enable me to delve deeper and deeper. I also discovered how broad this subject was, tying in with many areas of math in a fundamental way. That would allow me the room to branch out and explore in the way I so desired. I started to actively search for problems to work on. I also told Chern of my decision to focus on complex geometry rather than functional analysis, which I had thought was my preferred area when I first arrived at Berkeley.

Chern seemed okay with that plan, although he didn't express strong opinions one way or another. But in August 1970, after returning from a trip to Princeton, New Jersey, Chern suggested a rather dramatic course correction. He was excited by a conversation he'd just had with André Weil, a prominent mathematician based at the Institute for Advanced Study. Weil told Chern that mathematics had evolved to the point where proving the Riemann hypothesis, a classic problem in number theory, might be within reach. Riemann advanced his hypothesis in 1859 as a possible explanation for the distribution of prime numbers, which do not follow any obvious pattern. More than a century had passed, and no one had been able to prove that his idea was correct—or that it was incorrect—not even the great Riemann himself, who died at the age of thirty-nine.

Now Chern wanted me to take a crack at it. I was in need of a dissertation project, and he urged me to start working on this problem immediately. I had no doubt that the problem was plenty challenging—perhaps too challenging. But for some reason, it didn't move me. I'm just more excited by problems in geometry than in analytic number theory, which I think comes down to a matter of personal taste. When you start working on a big problem that could take years to solve, or at least make a dent in, you need to feel that excitement or you won't keep at it. In this case, my instincts might have served me well, as the Riemann hypothesis remains unsolved to this day.

Besides, by that time, I was already captivated by the Calabi conjecture. It's hard to say exactly why that is, though it's perhaps not that different from the fact that I was then a twenty-one-year-old guy who had seen a number of beautiful women, but only one of them—the one I'd spotted eight months before in Berkeley's math library—really got to me. I also had a gut-level, emotional response to the Calabi conjecture; that conjecture had gotten to me too. I knew it would be a long-term project, too much to take on for my PhD work, which meant I still needed a dissertation topic I could handle more expeditiously.

Fortunately, I got lucky less than a month later when Chern asked me to give a talk about the paper I had published in the *Annals* related to the work of Preissman. The talk went well enough, and afterwards Chern asked around to find out how good my paper really was. After some consultation, he decided it was good enough to be my PhD dissertation. I don't think he ever read the paper in any detail. It wasn't his field, as he didn't know much about group theory at all. In fact, very few geometers knew about group theory, although Joe Wolf was an exception, and he became a member of my dissertation committee. Lawson was on the committee too, as was the engineer Eugene Wong, at the request of Chern, in order to satisfy the requirement of having someone outside the math department participate.

Chern let me use a typewriter in his office so that I could type up my dissertation, which I did in early 1971. One fringe benefit of this was that Chern was so important in his field that geometers from all over the world sent him preprints of their papers. He let me read them, and when I found something interesting I could make a copy for myself. I discussed some of the most interesting papers I'd found during Chern's seminars. I've kept many of these papers and still find some of them interesting today.

Once I finished typing my dissertation and making the necessary copies, I was pretty much done. My work was accepted without my having to meet with my committee or answer questions or anything. It should have been a moment of great jubilation, as it's not every day that you qualify for a PhD. Yet a few things still dampened by joy. My relationship with Chern seemed strained; because he hadn't really advised me on this project, he thought that I didn't regard him as my advisor. Lawson had claimed me as his student, but he didn't teach me this subject either. I learned it on my own and never asked Lawson to be my advisor. As far as my dissertation was concerned, I learned the most from John Milnor's books and papers, although I did not meet Milnor himself until several years later.

I also felt somewhat disappointed that my graduate education would end so early, after just two years of study, because there was a lot more I still wanted to learn. But when your boss says you're ready for a promotion, it's best not to object too strenuously and argue that you should be held back. I went along with it all because there wasn't a good alternative and also because I wanted to be able to support my family in a more substantial way as soon as possible.

Backing up just a bit, I neglected to mention that in the fall of 1970, I got involved in something I hadn't planned on that continued well into the next year. I joined a group of overseas Chinese students in a protest known as the Diao Yu Tai movement. At issue was a group of eight small islands, none longer than a mile or two, which originally had belonged to

China. Japan took over the so-called Diao Yu Tai islands following its invasion of China in 1894, but after World War II the islands came under China's control, owing to their proximity to Taiwan. That changed in 1968 when Japan reclaimed the islands, with the backing of the United States, after subsea oil fields had been discovered in the vicinity.

Upset by an act of aggression aimed at weakening China, students in Berkeley and other U.S. cities voiced their disapproval. We were angered by Japan's militaristic tactics, as well as by America's support. We were angry at Taiwan, too, for its unwillingness to stand up in the face of imperialistic moves against China and for its efforts to actively suppress the Diao Yu Tai movement. The official paper of Taiwan said that students should be quiet and mind their own business, but that just made us madder. And louder.

Many of the Chinese students who were then in the United States, myself included, had never been involved in demonstrations before. But we modeled ourselves after the American students who were actively protesting the Vietnam War. We'd never done anything like this in Hong Kong and perhaps never would have. But the atmosphere was much different at Berkeley, where it seemed to be acceptable to take part in such an effort. I strongly felt that if we stood up for China, and China stood up for itself, we'd have more respect for our native country, and other countries would show it more respect too.

On April 9, 1971, we gathered for a demonstration in Portsmouth Square, a small park in the heart of San Francisco's Chinatown, with plans to march from there to the Japanese and Taiwanese consulates. Many of my friends had come too. I always carried a book at events like this, as typically a lot of time was spent standing around and waiting. On this occasion, I carried Morrey's book on differential equations but didn't have much of a chance to read it. Taiwan's government had hired a band of thugs to try to break up the protest. A friend of mine and fellow student, Yu, got knocked down in the skirmish, and many others were hurt. We proceeded with our march, as planned, but neither the Japanese nor Taiwanese consulates accepted our letters of protest. They ignored us, which motivated us to organize even more. Some students became full-time protesters. I didn't do that but still had less time to devote to math than during the previous year.

Around this time, Chern got sick and spent about a month in the hospital. I went with a group of Chinese students to visit him and was stunned by what he said. Chern was not happy with our political activism even though he, along with C. N. Yang and other dignitaries, had signed a letter published in the *New York Times* which made many of the same points that the student protesters were making. He advised us to stop these activities immediately. "The purpose of a man's life is either to gain fame or money," Chern said. "The student movement will bring you neither."

This was very different from what I thought our goals in mathematics were supposed to be—to seek, and hopefully find, the truth and beauty hidden within our chosen discipline—a lesson my father had impressed upon me, both implicitly and explicitly, throughout my childhood. The interaction with Chern reminded me of a classic Chinese essay my father asked me to memorize when I was around ten years old. The "Five-Willow Gentleman" was so named because he lived in an unfurnished shack—which offered scant shelter from the sun, wind, and rain—surrounded by five willow trees. Clad in rags and surviving on the most modest of resources, he was still content. He enjoyed reading so much that he often forgot to stop for meals. He cared not about his personal gains and losses, deriving satisfaction instead from writing down his thoughts and aspirations. The Five-Willow Gentleman was sustained by an intrinsic joy of study, not by the pursuit of fame or money, as Chern had put it.

I realized at that moment that, even though Chern and I did not always share the same values, I could still learn a lot from him. But I would have to keep his advice—as well as the advice given by anyone else—in perspective. I believe that he always had my best interests at heart. Ultimately, though, I would need to follow the dictates of my own heart.

But I was still a graduate student, of course. As my advisor, Chern had already done a lot for me and treated me quite well. I tried to do what was asked of me—partly out of gratitude and also because he knew the profession inside and out. And later that term, Chern asked me to teach a course on projective geometry, as he felt it would benefit me to have some teaching experience under my belt before I left graduate school.

About thirty students were in the class, and they seemed to be well behaved. T. Y. Lam had given me some lecture notes, which I found helpful in getting started. The trouble was, my accent was so strong that the students didn't understand what I was saying. One student complained to the department chair and dean about my teaching. Chern got nervous about this and asked Hung-Hsi Wu to see how I was doing. Wu thought my teaching was fine but agreed that my accent was a problem. The good news is that, over time, the students got used to my accent. Even the student who had complained to the chair and dean told them later that I was a good teacher. And from that point on, things went well.

I needed to start looking for a job, and Chern encouraged me to visit Stony Brook University on New York's Long Island because he felt it would be good for me to spend some time at another university. He got Jim Simons, the chair of Stony Brook's math department, to finance my visit. Lawson was visiting there too, as Stony Brook was then trying to recruit him.

I arrived at Stony Brook in March 1971, and Lawson let me sleep on his couch. However, that wasn't the best arrangement for his family, so I soon moved to the dormitories. I met a number of students in the dorms who were extremely frustrated by the political situation in Taiwan. At that time I also visited Columbia University because student leaders at Berkeley—who'd been involved in what started as the Diao Yu Tai movement and had since grown into a more generalized Chinese student political movement—asked me to meet with their New York counterparts. To my surprise, I heard that the Chinese student group in Columbia was angry at the Berkeley group for not consulting them before taking steps on their own. Their position struck me as so foolish, I barely knew what to say.

When I got back to Berkeley, our student movement continued, without the benefit of guidance from New York–based activists. But for me, finding a job had become a much higher priority, as my fellowship would not last much longer. I applied to six universities—the Institute for Advanced Study (IAS), Harvard, MIT, Princeton, Stony Brook, and Yale—and was lucky to get job offers from all of them. Harvard offered the most generous salary, $14,500 per year for an assistant professor position, which was decent money in those days. The other schools offered around $14,000 per year, except for IAS, which offered a one-year fellowship for just $6,400.

When I asked for Chern's help with this decision, he said, "Everyone should go to IAS at least once in his or her career, and you should too."

COMING TO AMERICA   73

Chern spent time at the Institute on numerous occasions, completing some of the best work of his career there between 1943 and 1945. I took his advice, without asking any more questions. Nor did I tell him about the comparatively low salary that IAS had put on the table—less than half of what the other schools had offered. Although money was a concern, I knew it wasn't everything, as the Five-Willow Gent had taught me. So I decided to take the long-term view: I'd go to Princeton for the year and try to get the most out of it. After that, I hoped I would land another position that would compensate me better.

But I still had one bit of pressing business to attend to before leaving Berkeley. I was determined to meet the woman who had made such a strong impression on me in the library a year and a half before. I'd seen her at seminars in the physics building, but we still hadn't exchanged a single word. I talked to a friend in the physics department who was from Hong Kong and finally learned her name, Yu-Yun, which sounded lovely to my ears—a reaction similar, perhaps, to how Tony felt upon hearing Maria's name for the first time in *West Side Story*—although I, fortunately, did not break into song.

My friend and I arranged a dinner between graduate students in physics and math, and it was his job to make sure that she attended. He seemed reluctant to do this, as I think he liked her too, but he came through for me. About three or four math students and an equal number of physics students attended that night, and we could all fit at the same table. I was finally "formally" introduced to Yu-Yun, which meant I could start dating her if she was amenable. We had only six weeks to get to know each other before we graduated and went our separate ways. I wanted to take full advantage of that time, although a few obstacles were strewn in our path.

Wu-Yi Hsiang, the young Berkeley professor whom Chern was fond of, invited me to his house for a lavish dinner. In the beginning, I had no idea that it was a setup and that Hsiang was trying to match me with his wife's relative. When I figured out what was going on, I discreetly told him I was interested in someone else. Hsiang was disappointed, which I suppose is a natural reaction in such circumstances.

Although I'm sure his intentions were good, situations like this aren't always so benign. A couple of years later I spoke with some Japanese mathematicians about Kunihiko Kodaira, who was the first mathematician from their country to win a Fields Medal. They told me that Kodaira asked one of their friends, a promising young student of his, to marry his daughter. The student agreed, becoming Kodaira's son-in-law, because he assumed that doing otherwise would be an affront to the great master.

Soon after I moved to the East Coast, I was invited to dinner by Wu-Yi's older brother, Wu-Chung Hsiang, who was a mathematics professor at Yale and then Princeton. At this dinner, he and his wife tried to set me up with one of his relatives. And like Wu-Yi, he too was disappointed when I told them I was already committed to another woman. On the one hand, it is certainly flattering to be thought of as a person to whom someone else would like to introduce a close relation in the hopes of sparking a possible romance. Nevertheless, the fact that things did not work out in these instances might have set me off on a bad footing with the Hsiang brothers, laying the groundwork for troubles down the road.

But back in June 1971, I got my PhD in math, while Yu-Yun got her PhD in physics. Chern wrote to CUHK's administration around that time, noting that I had a received a doctorate from Berkeley and should therefore be accorded an honorary degree from CUHK, given that I had never been issued an undergraduate degree. CUHK officials acceded to Chern's request, although they took their time about it, as I didn't get the honorary degree until almost a decade later, in 1980. A lot had happened by then—so much that I'd almost forgotten about Chern's letter.

During the summer of 1971, Yu-Yun and I spent as much time together as we could before and after graduation. Unfortunately, we were soon headed in opposite directions, at least in terms of geography. She was driving with her mother to San Diego, where she had lined up a postdoctoral position, and I was going to IAS, three thousand miles away, to begin my fellowship. We didn't know what the future would bring, though we promised to keep in touch. Given the uncertainties we then faced—two people, who'd known each other for only a short time, starting careers on opposite ends of the country—that was probably the best we could do.

CHAPTER FOUR

In the Foothills of Mount Calabi

WHEN I LEFT BERKELEY in 1971, at the age of twenty-two, my circumstances had suddenly changed in a significant way. For the first time since 1954, when I started going to school at the age of five, I was no longer a student. In other words, it was time for me to make my own way in the world and make my own decisions—rather than doing things simply to satisfy the expectations of my school or teachers or parents.

The place I would start doing that, the Institute for Advanced Study in Princeton, was tailor-made for this leg of my journey, and I was thankful to Chern for having directed me there, despite the financial sacrifice. IAS is the world-renowned institution at which Albert Einstein spent the last twenty-two years of his life. It sits at or near the pinnacle of the world's research centers. IAS was founded in 1930 so that scholars would be free to establish their own goals and do pretty much as they pleased, pursuing knowledge for its own sake, without regard to practical applications. In a 1939 essay that appeared in *Harper's Magazine,* the Institute's founding director, Abraham Flexner, wrote about the pursuit of seemingly "useless satisfactions" that can prove, unexpectedly, to be "the source from which undreamed-of utility is derived."

That philosophy held great appeal, as I already had a goal in mind that, on the face of it, appeared to have little or no practical utility. Yet I sensed that this work might ultimately afford some long-term benefit, not only to me but to others as well. I also knew I would need to acquire a good deal more knowledge before having a chance of transforming the apparently useless, to borrow from Flexner's phraseology, into the ultimately useful.

Whereas much of California is mountainous—with part of the California Pacific Coast Ranges running clear to the edge of the Berkeley campus—Princeton is decidedly flat. But even there, in New Jersey's fertile Inner Coastal Plain, with nary a hill in sight or even so much as a mound, I could still sense the presence of a mountain lurking nearby, one I someday hoped to climb. I called this peak Mount Calabi and knew it would involve a difficult first ascent. It would take me a while to identify a viable route and then prepare the tools needed to scale its rocky faces. The techniques I was contemplating involved novel ways of blending geometry and nonlinear partial differential equations, an approach that's now known as geometric analysis. As part of my toolkit, I would need to find solutions to a series of nonlinear equations that had never been solved before—a task that would take time, effort, and a good deal of luck. I wouldn't put myself onto Calabi's most treacherous pitches until those and other key elements were firmly in place. But I would not forget this mountain in the meantime because, for me, it was always there, hovering in the background and never far from my mind.

One of the great things about IAS was that a large group of us ate dinner together most nights, which meant there were always interesting people around with whom I could talk about math, as well as other topics of conversation should they arise. Let's just say there were no rules against shoptalk, and math did crop up from time to time.

Many of the folks had come for the year, like me, for the specific purpose of having such exchanges with other scholars and exploring ideas on their own that they were especially curious about. One of the people I most enjoyed speaking with was Nigel Hitchin, a young geometer just a couple of years older than me. Hitchin got his PhD at Oxford where he had been the assistant of Michael Atiyah, an internationally revered mathematician.

The Calabi conjecture was a popular topic of discussion between us. Calabi had proposed a systematic strategy for constructing a vast number of manifolds endowed with special geometrical properties of which we'd never seen a single example. Suppose a new planet were discovered, and a scientist soon came forth with a detailed plan for mining gold there—spelling out exactly where the mineral could be found and the quantities that could be extracted—before a single atom of that element had actually been seen on this world. Skepticism would be a reasonable response, which is why Hitchin and I, along with many others, considered Calabi's conjecture "too good to be true."

But it was still fun to think about his statement and speculate about the magical spaces it conjured up, while at the same time devising a hard-nosed plan for showing that it was wrong. Here's the line of attack I began to pursue: If the Calabi conjecture were true, then several of its corollaries—logical and inevitable consequences of the conjecture—must also be true. I just had to demonstrate that one of these corollaries was false, thereby establishing a "counterexample." I would then have proved, by extension, that the conjecture itself was false. Easier said than done, perhaps, but that at least appeared to be the simplest and most straightforward strategy. This approach is called "proof by contradiction." You assume a certain statement is true and then show that that assumption inevitably leads to something that is provably false—a contradiction, in other words.

A lot of great people were visiting IAS that year, including David Gieseker, a geometer who's been based at the University of California, Los Angeles, for the past several decades. Although Gieseker is just six years older than me, in Chinese culture we are trained to respect our teachers, and I paid close attention to his thoughts about geometry. I remembered our discussions well, and years later his ideas continued to influence my work. Chance conversations like this, I later realized, were a big part of the rationale for going to a place like IAS—and I suspect that other people have had the same experience.

It was especially great to interact with people from faraway lands. I had a lot of fun, for instance, hanging out with the Japanese mathematician Takuro Shintani, whose apartment was directly above mine. I learned number theory from Shintani, who later became known for the Shintani zeta function—a generalization of the Riemann zeta function that lay at the heart of the famous Riemann hypothesis—which Chern had tried to get me to work on for my PhD thesis.

Shintani was determined to learn how to drive a car while he was in Princeton but didn't get very far with that goal. He failed his driving test

An upright torus, or donut, offers a simple example of how Morse theory can be used to classify a manifold. According to the theory, a donut has four critical points. The uppermost point, or "maximum," has an index of 2 because there are two "independent" (and perpendicular) ways down from the top (see arrows). The next highest point, the upper saddle, has an index of 1 because there is just one independent way down. (It's not possible to travel downward on the surface in a perpendicular direction.) The next critical point (moving downward) is the lower saddle; it too has an index of 1, with just one independent way down. The lowest critical point, the minimum, has an index of 0; there is no way down from there because it is already located at the bottom. The torus can be defined topologically in terms of the indices of its critical points: 2, 1, 1, 0.

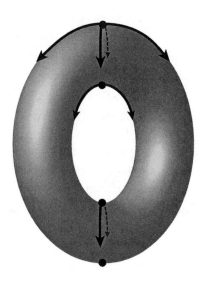

three times. Unfortunately, I was not a good enough driver to have been of any help, although I might have been able to show him what *not* to do. Nine years later, I was crushed when I heard the news that Shintani had taken his own life at the age of thirty-seven in the midst of a highly promising career. As we'd been out of touch, I have no idea what drove him to this desperate and tragic act. But I can say that in 1971, when Shintani arrived in Princeton, he was a vibrant and dynamic figure, as well as a pleasure to be around.

I also became friends with Ping-Fun Lam, a mathematician with a PhD from Yale who worked as a research assistant to Marston Morse, a prominent figure at IAS. Morse was famous for developing in the 1930s something called "Morse theory," which offered a new way to classify topological objects by focusing on a discrete number of so-called critical points, or transitional points, where the object's shape dramatically changes. I learned a lot about this idea from John Milnor's book *Morse Theory,* though I was surprised to discover that Morse hated the book and its title, preferring to call the subject "critical point theory." When Morse received a copy of Milnor's book, I was told, he tore it up and threw it in the garbage because he felt that he was the only one qualified to write such a tract. While that reaction seemed a bit overblown, I had no complaints about Morse personally, since he was very kind to me, as was his wife. I was careful not to upset our good relations by telling him how much I enjoyed Milnor's book and how much geometry I learned from it.

Apart from my usual activities at IAS—pursuing my research, attending seminars on a range of topics, and talking with other scholars in various settings—I also met with a group of people with ties to China who wanted to continue the student movement that was initially sparked by the dispute over the Diao Yu Tai islands but had since become a more diffuse enterprise. Although many of us were no longer students, we still wanted to keep that spark of discourse and activism alive.

Ping Sheng, a physicist then visiting IAS, joined us in these conversations, as did the mathematician Tzuong-Tsieng Moh, who had come to IAS from Purdue. Also in attendance was the aforementioned Wu-Chung Hsiang, who arrived at IAS about the same time I did. Hsiang seemed to have an uncanny knack for offending people, without necessarily trying to. I was the occasional recipient of his perhaps inadvertent skill, though I never let it get under my skin. Hsiang's wife, who was extremely nice, was constantly trying to save him from his missteps, although she was—in my opinion—only partially successful in those efforts.

One thing I liked about our meetings was that they gave me a regular chance to carry out conversations in Mandarin. We spent most of the time talking about Mao's book of quotations, also known as "The Little Red Book," which turned out to be, in fact, a rather big red book. I used this as an opportunity to learn more about Mao and his famous treatise, which I never had the chance to study in detail. But I eventually grew tired of these conversations, mainly because the guy who led our discussions, a physics graduate student named Bei-Lok Hu (who also had attended my high school in Hong Kong), treated Mao's book of quotations as if it were holy scripture.

With Hu at the helm, we could barely question anything in the book or probe it in a meaningful way, which left us with little of substance to talk about other than where to find a decent Chinese restaurant. There didn't seem to be any in Princeton, although one such eatery—located inside a supermarket—bragged about the fact that two Nobel laureates from China, T. D. Lee and C. N. Yang, regularly ate there. I still didn't love the food, despite the establishment's claim to fame.

We occasionally tried our luck in New York City for restaurants. Although we didn't find any sensational ones there either, among the hundreds, if not thousands, of Chinese dining spots in the metropolitan area, a few were good enough to satisfy our collective urge—at least for that week.

In July 1971, it was announced that President Richard Nixon would visit China in February 1972—the first time a U.S. president had been in the country since it came under Communist rule in 1949, the year I was born. It seemed that things were opening up in the People's Republic, and Sheng and Moh started talking about moving back to China. A lot of people urged Sheng and his wife not to return because they were expecting a new baby soon and would find it very difficult once they reached the mainland to buy the things they'd need—or had become accustomed to having in the United States. They were also warned, partly in jest, about the squat-style toilets that they'd have to adjust to in China.

For whatever reason, Sheng and his wife decided to stay in the United States, though they eventually moved to Hong Kong about two decades later. But Moh left for China in 1972, even though his friends and colleagues tried to talk him out of it. He quit his job at Purdue and left almost everything behind, including his car, which he tried to sell cheap but couldn't find any takers. But after only six months in China, Moh decided to return to the United States to look for a job. He was perhaps like many others who had quickly become disenchanted with the poor living conditions and paltry salaries in China. Unfortunately, he returned to a bad job market in the U.S. as well. Wu-Chung Hsiang said that he called Moh's former advisor at Purdue, Shreeram Abhyankar, and helped Moh get his job back. Luckily, Moh got his car back too, finding it just where he had left it. And, from what I heard, no one had stripped it down, siphoned the gas, or stolen the hubcaps.

I didn't spend all my time at IAS thinking about the Calabi conjecture, as I always like to have a number of projects going. During my year there, I started working on what are called "minimal surfaces," loosely speaking, the surface with the smallest possible area that is bounded by any small closed loop. If you take a round piece of wire and dip it into a

Joseph Plateau postulated that for any simple closed curve, one could find a "minimal surface"—a surface, in other words, of minimal area—bounded by that curve. The minimal surfaces spanning the curves in these three figures are called "Enneper surfaces," named after the German mathematician Alfred Enneper. (Based on original drawings by Francisco Martin, University of Granada.)

container full of soapy water, the bubble that forms will, in fact, be minimal, covering the smallest possible area and having nonpositive curvature.

I felt the subject had a lot of potential. In particular, I saw this as an area in which new methods from geometric analysis could yield a big payoff. At the time, most people were approaching minimal surface problems from the perspective of analysis. Geometers, meanwhile, were mainly focusing on the geometric aspects of these problems. It was as if the two groups were standing on opposite sides of a giant mountain, getting entirely different views. I was hoping to bring the two perspectives together. Although this had been tried before—on a limited, sporadic basis—I envisioned a synthesis in a large-scale, systematic way.

The field of minimal surfaces dated back at least to the 1700s with the work of the Italian mathematician Joseph-Louis Lagrange and to research in the 1800s by the Belgian physicist Joseph Plateau. After much experimentation with soap bubbles, Plateau postulated that for any simple closed curve, one can find a minimal surface bounded by that curve—a well-known conjecture that was not proved until 1930.

But there were still plenty of interesting problems to be solved in this area. Just as differential calculus can help identify the shortest path between two points on a given surface, it can also identify the smallest possible area that can stretch across a given loop. For that reason, I sensed that minimal surface problems constituted a good fit, and ripe target, for geometric analysis. Accordingly, I wrote several papers on the subject while I was at IAS.

An academic year at the Institute is short, ending in April, and thus goes by quickly. By December 1971, just a few months into my fellowship, I already had to start thinking about jobs for the following year. IAS offered to extend my stay for another year, which was unusual, and flattering, but I turned down that offer because I was worried about my visa status. I had a temporary F1 (nonimmigrant) visa, which is designed for students. In order to live and work in the United States permanently, I would need a green card. Without that, I could be kicked out of the country and sent back to Hong Kong. In terms of my research, that would be a big step backward for me, since comparatively little was going on in mathematics in Hong Kong as compared to the United States. Because of that, I had leapt at the opportunity to go to Berkeley two years earlier.

On the other hand, if I got a green card, I would be subject to the military draft—with the Vietnam War still going on. Paul Yang, a friend of mine who was working toward his math PhD at Berkeley, told me that because of my birthdate, I was likely to be drafted. I'm not sure whether he knew what he was talking about, but that prospect scared the heck out of me. I wanted no part of that war, which I had no stake in and which never made any sense to me. The student chants I'd heard countless times at Berkeley—"Hell no, we won't go!"—kept popping into my head. Although I hadn't marched with those students, I definitely embraced that sentiment.

Jim Simons, who was still in charge of the math program at Stony Brook, promised to take care of the visa situation for me, so I agreed to start there in 1972 as an assistant professor. As things turned out, this problem might have solved itself, because the United States ended the Vietnam draft in late 1972. But I had signed up for Stony Brook, and it was to Stony Brook I went—albeit after a slight detour.

When my appointment at IAS ended in April, I packed my stuff in a storage area reserved for the Fellows and then flew to California to spend time with Yu-Yun, who was still a postdoc at the University of California, San Diego (UCSD). I rented a hotel room nearby, as accommodations were cheap back then. I had fun hanging out with her, when she wasn't too busy with her research. She let me know, among other things, that my driving was deplorable, and she tried to improve my skills behind the wheel, though I didn't make much progress during that visit. When she was tied up with her work, I hung out at the UCSD math department, talking with the differential geometers Ted Frankel and Leon Green, among others. After a month or so in San Diego, I said goodbye to Yu-Yun once again. I then visited Chern and S. Y. Cheng in Berkeley before flying back to Princeton to gather my belongings and head to New York.

Ping-Fun Lam offered to drive me from IAS to Stony Brook, which is located along the northern shores of Long Island, some fifty-five miles east of New York City. I crammed my things into a small U-Haul trailer hitched to his car. Our route would take us right through Manhattan, and Lam said there was no way we could drive through the city without stopping in Chinatown. Negotiating that neighborhood's crowded streets with a trailer, finding a place to park, and then maneuvering into a tight space was not easy. But it was fun and ended up being a nice sendoff for me.

Tony Phillips, a mathematician who is still at Stony Brook, helped me find a car, which I thought I'd need given the school's out-of-the-way location. We drove a considerable distance, accompanied by Stony Brook mathematician Dennis Sullivan (who was then visiting from MIT), to find a used vehicle, a Volkswagen Squareback, for which the owner was asking $800. The car appeared to be in good shape, but that didn't last long. The next day I backed up in a parking lot and hit a pillar, smashing the back end of my vehicle. That's when I decided that Yu-Yun was right: I needed to become a better driver.

Not that there were many places worth going to in Stony Brook—a town that was in those days wholly lacking in culture though it had become something of a tourist destination. There was a mall and some shops and restaurants, but not much else. The good news is that with few distractions in the vicinity, I kept my focus, as usual, on mathematics.

I found a one-bedroom apartment not too far from campus. To save money, I rented out half of it to an undergraduate student from Hong Kong who slept on the couch. We started to cook together, but as in Berkeley, that experiment didn't last long. My cooking hadn't improved much over the past couple of years. It was so bad, in fact, that my roommate took over, which he probably viewed as a matter of survival.

The one thing I could cook was rice; I even had a rice cooker designed just for that task. I routinely made rice for dinner, which was part of my economizing in those days. Simons, who would eventually earn billions of dollars in hedge funds, was already doing well in the stock market. He occasionally poked fun at me for my frugal ways, though always in a good-natured manner. "There goes Yau," he would sometimes muse, "going home to eat his rice."

Late in 1972, Chern was on sabbatical at New York University's Courant Institute and came to Stony Brook to visit his friend C. N. Yang. A Nobel Prize winner, Yang was the first big shot to come to Stony Brook, hired to occupy the Einstein Chair. He arrived in 1967 and later became the first director of the newly opened Institute for Theoretical Physics. I'd known his name for most of my life but had never met Yang until I went to Stony Brook in 1972.

Chern was also in town to visit Simons, with whom he had collaborated on Chern-Simons theory—an influential theory in topology that is also relevant to quantum physics. I was supposed to drive Chern around and, thankfully, had upped my driving game somewhat by the time he arrived.

On Friday afternoons at 4 p.m., Yang gave a series of public lectures on fundamental physics. I attended those lectures regularly with the mathematician Howard Garland, a professor at Stony Brook who had received his PhD at Berkeley under Chern's tutelage a few years before me. Inspired by Yang's talks, Garland got very excited about physics, and he asked Chern whether he could switch fields. Chern told him it was too late; for better or worse, he was stuck with mathematics. Garland remained in the field, as Chern had advised, and became a good mathematician, choosing to work in an area between math and physics that made him happy.

The first class I taught at Stony Brook was elementary calculus, and I ran into the same difficulties I'd encountered as a graduate student at Berkeley. My accent was still very bad, and many students couldn't understand what I was saying. Enrollment in the course decreased dramatically after the first week of class—some students dropping out and others switching to other calculus sections. By the end, only four students out of more than a dozen remained. Nevertheless, those four students did extremely well on the final exam and were so happy they invited me to dinner to celebrate. Given the severe attrition, I guess I'd call the class a qualified success.

A young mathematician who gave a colloquium at Stony Brook that year, Reinhard Schultz, had just published a paper in an American Mathematical Society journal which showed that a ten-dimensional "exotic" sphere must possess a kind of "continuous" symmetry. An exotic sphere is topologically equivalent to a familiar, Euclidean sphere of the same dimensionality, though it lacks a more stringent criterion of equivalence known as "diffeomorphism." The term "continuous symmetry" is probably easiest to grasp by thinking of a circle. You can rotate a circle about its center by any amount—5 degrees, 37 degrees, or 489 degrees—and it still looks the same. That's what we mean by continuous symmetry, and that's the kind of symmetry Schultz identified in the exotic sphere. A square, by contrast, has "discrete symmetry"; it looks the same only when rotated by 90 degrees or multiples thereof. If you rotate a square by 45 degrees, for instance, it won't look the same. Instead, it will be tipped up on one corner, assuming the configuration of a baseball diamond.

Ordinarily I might not have paid much attention to Schultz's paper, except that Wu-Chung Hsiang had told me the year before at IAS that he'd found an example of a ten-dimensional exotic sphere that lacked circular symmetry. Although Hsiang's paper had not been published, he played up the accomplishment, telling me it was so monumental he needn't do anything else for the rest of the year. Schultz subsequently published two papers on the subject that established the existence of circular symmetry; his arguments were straightforward and looked correct to me. Hsiang might have agreed because, as far as I know, he never tried to publish his own paper.

The incident sparked my interest in the subject. I remembered that the first paper Hitchin published after getting his PhD also related to the ten-dimensional exotic sphere. I applied Hitchin's result to the problem Hsiang had worked on and found that while such a sphere could support continuous, circular symmetry, it could not support continuous, spherical symmetry. I discussed this result with Lawson, who was at Stony Brook then, and he made some important suggestions. We wrote a paper together that brought together all our findings. This paper was significant for me because it was the first time I had used geometry, specifically curvature, to prove something about differential topology. I later used the Calabi conjecture, a geometrical construct, to prove things in topology, but this paper with Lawson was my initial entrée into this area—spurred in part by Hsiang's broad assertions.

While at Stony Brook, I was still working hard on the Calabi conjecture, juggling it with my other research projects, particularly in the area of minimal surfaces, in addition to my teaching responsibilities. I teamed up, for instance, with the French mathematician Jean-Pierre Bourguignon, who was visiting Stony Brook from 1972 to 1973, the same year I was there. We tried out various approaches that might lead to the identification of a counterexample to the Calabi conjecture. Bear in mind that finding just one irrefutable counterexample would be tantamount to proving the conjecture wrong.

The work seemed to be going well. When I visited Chern in New York and told him I was close to finding a counterexample, he did not seem to know what I was talking about. When I explained things further, he was not visibly excited by this development, or potential development, at all. I was struck by how different our reactions were. When I had come across the Calabi conjecture for the first time, while sifting through documents in the Berkeley math library, I had been awestruck. The problem took hold of me, and I felt an overwhelming sense that this was something I had to prove. Regardless of whether I proved it right or proved it wrong, I was unable to turn away. Chern clearly didn't feel the same way. He had his own interests—things he cared about—but this proposition, for whatever reason, left him cold.

I, however, was guided in this effort by the conviction that the analytical techniques I'd learned from Morrey could be invaluable in tackling Calabi's problem. A primary focus during my previous year at IAS and the current year at Stony Brook was in developing "estimates" for solving highly nonlinear partial differential equations—the kind of equations in which the Calabi conjecture is written. It's a tricky business because solutions to these equations are not individual numbers but rather are functions—relations for which a given input yields a single output. In cases like this, involving fully nonlinear equations, we cannot expect to find an exact solution to an equation, such as a formula that can be written out explicitly, in full glorious detail. Our best hope, instead, is to find an approximate solution, or estimate, and then lay out a procedure for refining that estimate further until we can show it will ultimately converge on an actual solution.

I was starting to have some success in deriving estimates for a very important linear differential equation, and this represented a bit of a turning point in my career, as I have relied on this general approach pretty much ever since. I was proud of one estimate in particular and showed it later that year to Louis Nirenberg of the Courant Institute, a leading expert on partial differential equations. The fact that Nirenberg was not familiar with this estimate made me happy, because, given his exhaustive knowledge of the subject, it suggested that I'd done something new. The estimate turned out to be useful for me in solving a particular problem, but it also served a broader purpose, helping to propel me along this new direction that I've continued to follow.

During that same year at Stony Brook, I was also working hard on minimal surfaces, as mentioned. A two-part paper of mine on the subject had just been accepted by the *American Journal of Mathematics*. Although I didn't regard the paper as all that noteworthy, it attracted some notice nevertheless. In particular, it caught the attention of Robert Osserman, a Stanford mathematician who had made seminal contributions to the theory of minimal surfaces. Osserman was impressed enough by my paper, and some related preprints of mine, that he invited me to spend the next year, 1973–1974, at Stanford.

That was wonderful for me, not only because Stanford was (and still is) a great university but also because Yu-Yun, whom I still had designs on, had just lined up a postdoctoral fellowship at Stanford that would start in the fall of 1973. We had spent two years on opposite ends of the country—she in San Diego, and me in New Jersey and New York. We'd finally have the chance to be on the same coast and, in fact, on the same university campus. I found the prospect exciting, though I knew it would put our relationship to a test.

The timing of the Stanford appointment was convenient for another reason. I was already planning to attend a major conference on differential geometry that would be held at Stanford from July 30 to August 17, 1973. Many of the big players from all over the world would make an appearance, and I didn't want to miss it.

Cars are a big part of the California culture, as several Beach Boys hits would attest: "Little Deuce Coupe," "I Get Around," and "In My Car." I figured I'd better have one out there, which meant I needed to drive my VW Squareback across the country—a daunting proposition given my spotty track record with regard to automobiles. Fortunately, Wen Chiao Hsiang, a graduate student of Jim Simons who was interested in differential geometry, wanted to attend the conference too and volunteered to help with the driving. Having no experience with long-distance driving,

I went to AAA to get maps, driver's insurance, and traveler's checks. I lost all the traveler's checks within a few days, but AAA was kind enough to replace them (which is, after all, the main point of getting traveler's checks in the first place). Hsiang and I planned a route that would take us through Yellowstone and other scenic stops along the way. We allowed ourselves almost two weeks to get to California, leaving plenty of time for sightseeing.

We set off sometime in May to "see the USA," as Dinah Shore used to sing in that old Chevrolet jingle. For me, this trip was the first time I was able to appreciate the natural beauty of the United States. I also got a sense, during our three-thousand-mile drive, of just how big this country is. We made it in one piece without much difficulty other than a flat tire, which Hsiang helped me replace somewhere in the middle of nowhere.

Before getting to Stanford, we stopped off at Berkeley. I visited S. Y. Cheng, who had recently gotten married to the young woman who lived next door to us when we were roommates on Euclid Street. The next morning, on the way to meeting Hung-Hsi Wu in the math department, Wu-Yi Hsiang urged me to come to his office for a chat. Because it was up his alley, I told Hsiang about the recent paper I'd done using partial differential equations from geometry to solve a problem in topology. Hsiang dismissed the work as trivial, insisting he could prove the same thing with topology alone. In so doing, he was exhibiting the same bias that had surfaced at a Berkeley seminar, which I led at Chern's request, on the subject of solving problems in pure topology with differential geometry. At some point, Hsiang stormed out of the room, in front of Chern and many others, after expressing his opinion that topologists did not need help from geometers in solving problems in topology.

While we were in his office, Hsiang went to the blackboard to sketch out his approach—one that would supposedly show why it was unnecessary to apply geometry to the topological problem at hand. But after about an hour, he still hadn't managed to persuade me of his contention. Hsiang left the room abruptly, saying he had to go to the bathroom. I waited for a while, but then left to have lunch with Wu and never heard more from Hsiang on this point.

I arrived at Stanford in June, a month before the big conference was set to begin. I rented an apartment on University Avenue, the servant's quarters of a big house, which had just one deficiency: It had no kitchen.

When Chern and his wife visited a short while later, Mrs. Chern made fun of me because the hotplate, which I used for cooking rice, was right next to the bathroom. "What you're cooking will come out right here," she laughed, pointing to the toilet.

I still liked the apartment—that drawback duly noted—and I soon became friends with the Chinese couple who lived next door with their daughter and son. The daughter, coincidentally, ended up marrying my friend Ronnie Chan, a Hong Kong businessman who later became a generous donor to Harvard University and various mathematical endeavors in Asia that I've been associated with.

Osserman gave me an office on the second floor of the math building. It was a small room but well situated, as I was right next door to Leon Simon, an Australian mathematician who soon became a great colleague and friend. Simon had gotten his PhD just two years before from the University of Adelaide, and I think the Stanford department chair, David Gilbarg, showed great judgment in hiring someone who'd come from such a remote and not terribly prestigious institution. Simon and I were joint advisors to a new graduate student, Rick Schoen, who was just a year younger than me. I guess I was in the right place at the right time, as I've greatly valued my association with Simon and Schoen, both of whom are truly original mathematicians. The three of us worked together well; we learned from each other and complemented each other's strengths. I'm convinced that having this core group in such close proximity for several years created a kind of critical mass that went far toward establishing geometric analysis as an actual field rather than just a vague notion I'd been toying around with on my own.

I was looking forward to the Stanford conference because it was going to be a truly international affair; practically everyone who'd done something related to differential geometry would be there. Chern and Osserman asked me to give two talks, both related to work I'd done on minimal surfaces at IAS and Stony Brook. Lawson was also going to talk about our joint project on exotic spheres. So I had a lot to think about before the conference got under way—preparing my own talks and also wondering about what Lawson might say regarding our collective effort.

One lecture I attended at the event, given by the University of Chicago physicist Robert Geroch, made a deep impression on me. Geroch spoke of the positive mass conjecture, a statement from general relativity holding that the total mass or energy of any isolated system—including the universe itself—must be positive. Physicists, by and large, believed this conjecture was correct, but they'd been unable to verify it. Geroch thought that geometers might be well positioned to prove it. Although most geometers were not terribly interested in physics at that time, I was intrigued. I decided that Geroch's proposition was not too farfetched, given that the conjecture could be recast in strictly geometric terms: If the matter density of an isolated physical system is positive, the total mass of this physical system due to gravity must be positive too. A positive matter density implies positive average curvature, and curvature was something geometers spent a lot of time thinking about. I personally never tired of the subject and immediately began to consider how techniques from minimal surface theory might be brought to bear on this situation. I kept the problem in mind until Schoen and I had a chance to pursue it several years later.

But something else happened at the conference that soon took over my life. I spoke with a bunch of people—including Eugenio Calabi, Robert Greene (of UCLA), Louis Nirenberg, and Hung-Hsi Wu—about differential geometry, mentioning ideas that might be useful in attacking the Calabi conjecture. At some point during these conversations, I mentioned that I'd come up with a seemingly robust counterexample or two. Word got around, and I was asked to give an informal presentation on that subject after dinner one night. Thirty or so people showed up, including Calabi and some of his University of Pennsylvania colleagues. There seemed to be a lot of anticipation in the room, which made me a bit nervous, perhaps, though I still felt confident in the material. I talked for about an hour, and it all went smoothly. No one spotted any flaws in my argument or challenged any of my assertions. And I comfortably dispatched any question thrown my way.

By the end of this session, most people left the room with the sense that I had proved Calabi wrong. Both Calabi and Chern let me know that they thought I'd put forth a good counterexample. Rather than appearing upset by this outcome, Calabi instead seemed relieved that, after almost twenty years of uncertainty, the matter was finally resolved. Chern, meanwhile, told me that my presentation had been the high point of the entire conference, which is always nice to hear.

The conference ended in mid-August, giving me a few weeks to get settled at Stanford before the fall term began. I continued my work with Simon and Schoen, while also getting to know other members of the department. I met the algebraic geometer Bruce Bennett—a former student of Heisuke Hironaka, a Fields Medal winner from Japan—who was a fine mathematician in his own right. A large and extremely muscular guy, Bennett once tore apart a door in a public restroom, not out of any destructive impulse but simply because he was in a hurry. Garo Kiremidjian, another junior faculty member like me, worked on complex manifolds, just as I did, and we had many fruitful discussions.

I also spent a fair amount of time with Kai Lai Chung, an expert in probability theory who was born in Shanghai. Chung, who was about thirty years older than me, liked to take walks in Palo Alto's parks. I joined him on a number of strolls, during which he liked to recount stories and anecdotes about older mathematicians like Chern and Hua between whom a famous rivalry had developed. I was a highly receptive audience, so we made a good pair.

In these tales, Chung was always complimentary to Hua, having studied with him years earlier in China, but Chung never had anything good to say about Chern. Through our talks, and from my subsequent digging around, I learned some of the reasons why Chern and Hua didn't get along—a situation that had adverse consequences for the entire Chinese mathematics community, as well as for me personally.

As Chung recounted the story, Hua was considered a genius because he had solved some big mathematical problems, despite the fact that he grew up in a poor family and had to make it on his own with an extremely limited education. Chern ultimately made far bigger contributions to the field, but those came somewhat later. Chern didn't face the same financial struggles as Hua because his father was a judge, whereas Hua's father was a shopkeeper and none too prosperous. In 1941, the Chinese government gave Hua its first-ever national scientific prize, a prestigious award something like the National Medal of Science that the United States started handing out a couple of decades later. I imagine this came as a blow to Chern, who happened to be living with Hua at the time. Chern's resentment might have increased over the years because he was never accorded this same honor, even though Chung—who was telling me this story and never came close to rivaling Chern in stature—later won a silver medal.

The rift between Chern and Hua might have started because of this perceived slight, growing wider as the years passed. It doesn't take much, I've noticed, for feuds to begin, but it can take a lot to end them. Sometimes they don't end until the main players have passed on and no one is left to fight.

Chung was an unusual guy who didn't get along well with other people in the department. He and Sam Karlin, who also worked in probability, didn't even talk to each other. Even though I was on the faculty, I often sat in on classes. I sat in on Chung's probability class in which he discussed Brownian motion—a phenomenon, stemming from the constant motion of atoms, that was first explained mathematically by Einstein.

At the end of the term after finals, Chung handed out a special problem for extra credit that was quite challenging. Several students put a lot of effort into it. While working through the problem, the students needed a reference to a topological statement they felt must be true. The Harvard mathematician Andy Gleason, who was visiting Stanford at the time, directed them to a paper by Kazimierz Kuratowski, which had just what they were looking for. Shortly thereafter, the students laid out their argument to Chung. He stopped them when they reached the point of applying the Kuratowski result. One student mentioned that Gleason had apprised them of it. "Just as I thought," said Chung, or words to that effect. He then promptly walked out of the room, even though the students were in the middle of their presentation.

I witnessed the whole episode, aghast. I could not believe that Chung would treat his students so callously (even though he had been very nice to me during my time at Stanford). Maury Bramson, who was a Stanford math graduate student at the time, told me that Chung's unpleasantness was a major factor in his decision to leave Stanford and finish his PhD at Cornell.

I often went out to eat with Bramson and some of the other younger guys in the department. (I suppose I was a "younger guy" too, despite my faculty status; as a result, I probably spent more time interacting with the graduate students than most of the professors in the department.) Our favorite spot was Moon Palace, but on Saturdays we tended to go to a place called Peking Gardens, I believe, which had an all-you-can-eat luncheon buffet. On one of these outings, I remember Bramson polishing off five full plates, and then he didn't eat again for two days (though Bramson recalled that he "could not go a full day without eating"). The owner of the restaurant was so happy that someone liked her food so much that she didn't charge him.

Once I was sitting in my office and heard someone outside my door speaking in perfect Cantonese. I assumed it was S. Y. Cheng, who was then visiting Stanford, but it turned out to be a graduate student named David Bailey—a Mormon who'd just gotten his bachelor's degree from Brigham Young University. Bailey needed to satisfy his foreign language requirement. I tried to trip him up during his preparations by giving him a difficult bit of text to translate. It was written in "simplified" rather than traditional Chinese, but in this case simplified is actually a good deal harder than the standard text. I assumed that Bailey would struggle with the task, but he did a beautiful job.

On another occasion, I overheard Bailey's conversation with an older graduate student as they sat on a sofa outside my office. Bailey was working on a mathematical problem and wanted to learn about the process of getting a paper published in a refereed journal from someone who knew the ropes. "Doing mathematics is just like ____ing a girl," the older and wiser student told him. "The first time you may have some problems, but the next time it tends to go more smoothly." That's not how I would have put it, though the advice may have helped, as Bailey went on to have a fine career in mathematics before switching to computer science, where he achieved similar success.

As for me, life at Stanford was shaping up well. I met with Yu-Yun as often as I could, though both of us were busy, starting up in new positions. Consequently, we weren't together much. But we've always been rather independent, even after four decades of marriage. Back then, however, our situation seemed a bit precarious. I thought it best to keep our relationship separate from the math world until I had a better sense of where the two of us stood. As a result, very few of my colleagues in those days knew about Yu-Yun or my feelings for her. Perhaps I couldn't express those feelings as easily to her as I could write down a mathematical equation, which could explain why it took many years for our relationship to gel. That, in turn, might have been an occupational hazard of my trade, though more likely it was a reflection of the kind of people attracted to that trade—people, like me, who are generally more fluent in numbers than words.

Meanwhile, most people in Stanford's math department had gone out of their way to make me feel welcome. And for the first time ever, I had a secretary—a nice Chinese woman named Frances Mak—to type up my papers, which really helped my productivity. I still spent most of my time doing math, but there were a lot of agreeable ways to take breaks. A walk on the campus, through its perfectly manicured grounds, was always a pleasant diversion. One could see palm trees and hills in most directions, and the buildings—in the Spanish colonial style, with stucco exterior walls and red-tiled roofs—never failed to impress. Sometimes a group of us would toss a Frisbee around or play ping-pong at a table near my office. I never had to work too hard to find someone to go out with for Chinese food, and the dining choices were better than in Princeton.

All in all, I was happy with Stanford, and Stanford, apparently, was happy with me. Later in the fall, after I'd been at there a couple of months, I met with Osserman and the department chair, Ralph Phillips, who encouraged me to stick around. They offered me an associate professorship without tenure, agreeing to write a letter stipulating that I would be granted tenure a year later.

At around the same time, both Johns Hopkins and Cornell offered me associate professor positions. I don't know how they heard about me, but I figured that Chern had something to do with it. He was good friends with Wei-Liang Chow, an important mathematician from Shanghai who was on the Johns Hopkins faculty. The Johns Hopkins offer was not so attractive, in part because I'd heard that associate professor appointments there typically do not lead to tenured positions. Chern had also been the advisor of Hsien Chung Wang, originally from Beijing, who was then teaching at Cornell. One big supposed inducement for me at Cornell was that Wang had volunteered to help me find a Chinese woman to marry. But that wasn't a great incentive for me, as I already had someone else in mind for the job (though, frankly, it was hard to tell how much progress I was making on that front).

Although I'd been at Stanford only a few months, my situation seemed secure. There wasn't much pressure on me at the moment, other than having to choose between several attractive job offers. But I was just getting started at Stanford, and the thought of moving again so soon was not very tempting. I preferred to stay where I was and enjoy the California lifestyle, maybe even relax a bit—although the word "relax" was barely part of my vocabulary. Nevertheless, my stress level was perhaps at an all-time low.

Around that same time, well into the fall of 1973, I received a short though polite note from Calabi. He'd been thinking about my presentation in August and upon reflection had found some aspects of it puzzling. He asked me to sketch things out on paper so he could better understand my argument. I hadn't gotten around to doing that yet, but Calabi was right: I needed to take things to the next level—or, to put it in other terms, get back on that "mountain." For if my counterexample truly held up, this conjecture—which had been kicking around for nearly two decades—was dead in the water. Maybe I hadn't taken the next step because I liked the conjecture and wasn't ready to write its obituary.

For me, Calabi's letter served as a wake-up call. For the next two weeks, I put everything else aside and worked almost nonstop, barely taking time off to sleep or eat. I picked what seemed to be my strongest counterexample and began to construct the argument, but it didn't hold up under intense scrutiny. At the last minute, when I was about to put the finishing touches on it, the argument would suddenly fall apart. When I went through the other possible counterexamples I'd been considering, one by one, they fell apart too. It was frustrating and maddening, putting me into an agitated state that made it difficult for me to rest or think of anything else. Instead, I kept going feverishly, compulsively, unable to stop. But the longer I stayed at it, the more I realized that my strategy was doomed to failure.

I'd spent two weeks, nearly killing myself in the process, trying to prove the Calabi conjecture wrong. Now I had to face the prospect that this conjecture—which Hitchin and I and many other colleagues considered "too good to be true"—might be true after all. And as time went on, I became convinced, in fact, that it *must* be true. So I now had to reverse my course 180 degrees and pour my efforts into proving that Calabi had been right all along. I didn't know exactly how to go about doing that, but one thing was apparent from the outset: It wasn't going to be easy.

CHAPTER FIVE

The March to the Summit

IN 1746, Gaspard Monge was born in the French town of Beaune, which is located near Dijon in the heart of Burgundy wine country. The son of a peddler, Monge showed a knack in his youth for architectural drawings. While still a teenager, he drew a large-scale, incredibly detailed plan of his hometown, which caught the attention of a military officer who helped Monge gain entrance to a military school in northern France. Owing to his common origins, Monge was not admitted to the school proper, which was reserved solely for aristocrats, but he was allowed to study drafting and surveying in a separate part of the facility. This arrangement was not entirely satisfactory to him, as he longed for a position that would enable him to utilize his talents more fully.

Monge got his chance a year or so later: He was asked to determine the best place to position guns at a proposed fortress so its occupants would be well protected from enemy fire. He used geometric techniques, which he developed on his own, to solve the problem, completing the task so quickly as to arouse suspicions in some quarters. Nevertheless, his mathematical skills could not be denied, and Monge was finally given the opportunity to cultivate them.

He started teaching physics and mathematics in 1768, conducting research on partial differential equations as well as applications of calculus to geometry. In the 1780s, after assuming a mathematics post in Paris,

Monge began studying nonlinear partial differential equations of a special kind, which were eventually called Monge-Ampère equations—a designation that presumably reflects some modifications made several decades later by the French scientist André-Marie Ampère, who was best known for his contributions to the theory of electromagnetism and after whom the unit of electric current, the ampere, was named. (I say "presumably" because I'm not aware of any actual contributions that Ampère made to this subject, though he may have done so. On the other hand, sometimes a name is stuck on an equation for no apparent reason.)

Monge's story shows, first of all, that a career in mathematics can be launched in indirect and unexpected ways, although some underlying aptitude for the discipline is always helpful. But my principal motivation for bringing up this anecdote stems from the fact that the Calabi conjecture can be expressed in terms of a Monge-Ampère equation. This kind of equation is nonlinear, as mentioned before; has at least two independent variables; and is also "complex," meaning that it involves complex numbers. The challenge, from my end, was that no one had ever solved a complex Monge-Ampère equation before, except in the simplest, one-dimensional case. But in tackling the Calabi conjecture, the higher-dimensional form of these equations, which had so far defied resolution, would have to be solved too. This was the big stumbling block and the reason why little progress had been made on this problem twenty years after Calabi first raised it.

I started working on Monge-Ampère equations at Stanford during the 1973–1974 school year, about two centuries after Monge began formulating his ideas on the subject, although I was lucky to have some mathematical tools at my disposal—including a few I devised myself—that he probably could not have imagined. I first looked at Monge-Ampère equations defined over real numbers, which relate to the curvature of a surface. The real equations are easier to manage than the complex variety, and I enlisted the help of my friend S. Y. Cheng, who often came down from Berkeley to visit. Our plan was to build up some mastery over the real equations before taking on the more demanding complex ones.

Fortunately, Cheng and I had some success. We solved a Monge-Ampère–type equation that appeared in the famous Minkowski problem. That problem, in very simplified terms, relates to showing whether an object with a distinct kind of curvature could exist. You can probably see why this problem appealed to me, as I had been intrigued by the connection between geometry and partial differential equations ever since taking my first course with Morrey four years earlier. This was, indeed, the principal thrust of the emerging field of geometric analysis, whose development I was doing my best to foster, while encouraging colleagues like Cheng, Schoen, and Simon to join me in this effort.

The general strategy for solving equations of this sort, as discussed in the previous chapter, involves coming up with a series of approximate solutions and narrowing down the range until it can be shown that this process will eventually converge on an actual solution. My hope was that I could eventually do the same with the complex Monge-Ampère equation that encapsulated the Calabi conjecture. Proving that a solution to this equation existed would be equivalent to proving the existence of the extraordinary geometric spaces postulated by Calabi—spaces endowed with a special symmetry and distinct curvature that also satisfy the Einstein equations.

In the spring of 1974, Chern invited me to Berkeley to give a talk. The Russian-born mathematician Mikhail Gromov was then visiting Berkeley for the first time, where he was treated like royalty owing to his reputation as one of the world's great geometers. I'd had a not-so-wonderful encounter with Gromov about six months earlier. At that time, I had used geometric analysis to prove that a certain space had an infinite volume. Gromov claimed that my proof must be wrong although I'm not sure he really understood the approach I had taken. In the end, this result has held up quite well.

I was discussing a different subject at Berkeley, which had to do with the "spectrum" of a geometric space—the resonant, vibrational frequencies produced by deforming a space, which are similar, in principle, to the characteristic frequencies produced when a drumhead is struck and its surface thereby deformed. Gromov took issue with me once again, announcing in the middle of the talk that he considered my line of attack to be fundamentally unsound. The proof I was then discussing, as with the previous proof we had debated about, relied heavily on nonlinear partial differential equations—an area that Gromov did not specialize in. It's possible that he just did not understand my proof. But rather than asking me to explain the argument, he instead contended that I didn't know what I was talking about.

That appeared to be his modus operandi, acting as if I were a lowly student who hadn't done his homework properly. He took up much of the time allotted to me in the seminar to voice his skepticism about my work. What it came down to, as best I could figure, was that he did not regard geometric analysis as a worthy pursuit. Any theorem in geometry, he insisted, must be proved by geometric means. I, of course, had a different opinion, and the whole premise of geometric analysis depended on the strength and validity of that conviction.

The seminar was not a great success and hardly could have been, owing to Gromov's loud and frequent interruptions. But afterwards I spent hours explaining to him my new result and my earlier proof on infinite volume, answering one doubting question after another. I finally showed him how to translate my analytic method into strictly geometric terms, at which point he relented, giving tacit approval to my results.

I later had similar though much more cordial interactions with Bill Thurston, a fellow graduate student of mine at Berkeley who had since made a name for himself in geometry and topology. Thurston's approach to geometry was somewhat akin to building geometric spaces, or manifolds, out of small pieces like Lego blocks and in that way delineating their inner constitution. I took almost the opposite strategy, using differential equations to get a handle on both the underlying structure of objects and their overall topology. The two philosophies were quite different, although both have ultimately proved successful. I should stress that Thurston was a really deep thinker and a true original. While he did not always go into great detail in some of his arguments, the ideas he formulated have had a profound and lasting influence on the field.

I learned a valuable lesson from my exchanges with Gromov and Thurston and other conversations of a similar nature: I would have to overcome a lot of resistance among some mainstream geometers and topologists before the tools of geometric analysis were widely embraced. But I suppose that is the way when all new techniques, especially dramatically different ones, are brought to the fore. Such a guarded response can serve as a helpful precaution, but it can also hold back progress in a field.

I didn't let any of the skepticism dampen my enthusiasm for this line of work, which seemed to be progressing well enough. Nevertheless, I suffered a personal setback in June 1974. Yu-Yun, who'd been a postdoc at Stanford, got another postdoctoral position at the Princeton Plasma

Physics Laboratory—a U.S. Department of Energy lab located on the Princeton University campus. This was an excellent appointment for her, and ordinarily I would have been thrilled. But it meant that we would soon be on opposite ends of the country again. She left shortly thereafter, driving to Princeton with her mother.

A surprise visit by Tat Chui, my old pal from Hong Kong, and his girlfriend served as a welcome distraction. It turns out that his girlfriend, who was soon heading back to Hong Kong, was on the verge of breaking up with him. We decided on the spur of the moment to go to Yosemite National Park, heading out that evening in my car and arriving in the mountains late at night. This brief outing turned out to be just the thing for all of us. Breathtaking views from lofty peaks have a way of magically taking you outside of your own head, offering a new and broader perspective on the world. We had such a good time during this trip that Tat and his girlfriend decided to get married.

Although I was happy for them, I was on my own—at least for the time being. And I did what I always did in such situations: I threw myself into my work. I was used to working extremely long hours—often going late into the night and occasionally falling asleep at my desk—a lifestyle that was, admittedly, not ideal for nurturing a relationship. But now that I was alone, I had no shortage of mathematical projects with which to fill my time and thoughts, especially the Calabi conjecture, in which I was deeply and irretrievably immersed.

Before taking on the complex Monge-Ampère equation that lies at the heart of that conjecture, Cheng and I felt that some additional preliminary work was still needed. In 1974, we started to work on the so-called Dirichlet problem, named after the German mathematician Peter Gustave Lejeune Dirichlet, without realizing that Eugenio Calabi and Louis Nirenberg were working on it at the same time. Classified as a "boundary value" problem, the basic idea can be summed up as follows: Just as solutions to simple equations might consist of a set of points that define, for example, a circle or parabola, the solution to more complicated differential equations could define an entire surface. If all you know is the boundary of that surface, the Dirichlet problem asks, can you use that knowledge to find all the interior points that make up the surface and at the same time satisfy the equation at hand? The standard procedure, as with the previous exercises, involves making a series of estimates, or ap-

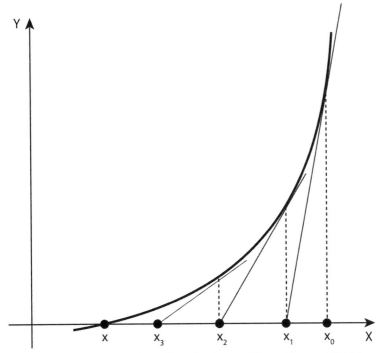

While the estimation methods used to solve the Dirichlet problem and Calabi conjecture are too complicated to present here, we can instead provide an example of a very simple approximation technique called "Newton's method." To find the roots or zeroes of a function whose output is real numbers, places where a particular curve crosses the X axis, we start off with our best guess—a point we'll call X_0. We then take the tangent of the curve at X_0 and see where the tangent line crosses the X axis (at a point we'll call X_1). As we continue this process, assuming our initial guess is not too far off, we'll come closer and closer to the true answer, point X. (Based on original drawing by Xianfeng [David] Gu and Xiaotian [Tim] Yin.)

proximations, that collectively allow you to home in on a function that solves the specified partial differential equation.

Nirenberg was scheduled to give a plenary talk at the International Congress of Mathematicians, which would convene in Vancouver in August 1974. In this talk, he intended to present the solution to the Dirichlet problem based on the work he and Calabi had done. Chern, however, told us that Calabi and Nirenberg had subsequently found a mistake in the preprint they had circulated on the subject that invalidated their solution and left the problem up for grabs.

I told Chern of my confidence that Cheng and I could solve this problem. Since Nirenberg was coming to Berkeley sometime in the spring of 1974, Chern arranged for the four of us to meet for lunch at Louis' Restaurant in San Francisco, located on the beach not far from the Golden Gate Bridge. Cheng and I went through our proof carefully the night before to make sure there was no problem, as Nirenberg was a leading authority on partial differential equations and we didn't want to embarrass ourselves. We uncovered a mistake in our argument, but after further work, we felt that we had fixed it by 2 a.m. Before lunch, we described our solution to Nirenberg, who thought it sounded reasonable. Cheng and I were pleased but later uncovered more mistakes when we went through the proof again that night. The errors were frustrating, but they helped us learn how to deal with this kind of equation. Six months later, we figured out how to fix them and ultimately solve a weaker version of the Dirichlet problem. Nirenberg and others solved a stronger version of the problem about ten years later.

But the San Francisco meeting with Nirenberg proved helpful to me in another way. Earlier in the year, Robert Osserman had nominated me to become a Sloan Fellow—an honor for a young assistant professor—which would allow me to spend a year wherever I wanted, with my salary paid in full by the Sloan Foundation. I initially thought I'd use the fellowship to go to Princeton, where I'd have the chance to see Yu-Yun, with the hope that she and I could renew our relationship—and maybe more.

I wrote to Wu-Chung Hsiang, who had moved to Princeton from Yale, to see whether I could spend at least half of my fellowship year in Princeton. A few days later he told me the math department had no office space available. Now that I have some familiarity with these things, I suspect that space for me at Princeton could have been found if Hsiang and others in the department wanted me there. I realized, belatedly, that I might have had better luck by writing to the department chair. Instead, I went through an acquaintance—a possible mistake given that the person I knew may not have been all that fond of me. (The situation was reversed a couple of years later, ironically, when the Princeton math chair asked Hsiang to call me to offer a professorship. On that occasion, I turned him down, though I was not at all acting in spite; I just wasn't ready to move to Princeton then.)

Fortunately, Nirenberg said without hesitation that I should spend the fall of 1975 at New York University's Courant Institute, which was not too far from Princeton. Nirenberg was enthusiastic about my coming to Manhattan, and by the end of lunch it was essentially a done deal.

Once the Dirichlet problem (in its weaker form) was behind me, I turned my attention toward the Calabi conjecture. My strategy was pretty straightforward: I planned to take everything I had learned from the work on real Monge-Ampère equations and transfer that, to the extent possible, to the complex case. Cheng, who had little expertise in complex geometry, bowed out at this stage to undertake projects closer to his interests and expertise. Coincidentally, he was going to be at Courant too, which would enable us to spend time together socially, as well as continue our joint efforts to build the foundations of geometric analysis.

Going to New York, which I did in August 1975, was beneficial to me for another reason. Thanks to my Sloan Fellowship, I had no teaching responsibilities and was free to work on the Calabi conjecture, or any other mathematical problem, as much as my time and energy would allow.

I intended to take full advantage of that opportunity, though my first order of business was finding a place to live in Manhattan, where accommodations were very expensive. Studio apartments were going for $200 per month and up, which was more than I wanted to spend. Fortunately, I caught a break from Jürgen Moser—a former Courant director and friend of Chern's—who had access to a friend's rent-controlled apartment on Spring Street, near New York University. The rent was just $50 a month, a fabulous deal. Moser was not supposed to rent out the apartment because it was not leased in his name, so I was told to avoid conversations with the owner who, coincidentally, was Chinese. He didn't speak English, though he spoke Cantonese, which I knew very well, but I had to pretend I didn't understand a word he said. And if anything went wrong with the apartment, I had to tell Moser, who would take care of the problem himself. Given all that, it was rather amazing, and incredibly generous, for Moser to have done this for me, a virtual stranger.

While Courant provided a wonderful place for me to work, my main purpose in going there, as opposed to any other destination I might have chosen for the fellowship, was to be close to Yu-Yun. We had had little contact since she had left Stanford almost fifteen months earlier. But if I were going to date her, and take her places outside of the campus, I would need a car. Unfortunately, I didn't have a credit card then and was unable to rent a car in New York City without one. I got Stanford to write a letter on my behalf explaining I was on the faculty there and temporarily at Courant, but that held no sway over the car rental agencies.

I started to panic because, without a car, my whole plan of spending time with Yu-Yun, which had taken me to the East Coast in the first place, would be seriously jeopardized. Luckily, I came across a high school friend who was in New York working as a travel agent. He told me about a low-end, "rent-a-wreck" kind of establishment that would loan me a vehicle after I put down a hefty cash deposit. The car I got was in barely drivable condition and nothing great to look at, but it would have to do, given the limited options available to me.

Despite its shabby appearance, the car was good enough to get me to Princeton, and I visited Yu-Yun there when I could. She was tied up with her research, just as I was preoccupied with my work on the Calabi conjecture, but it felt like I was making steady progress on the latter. Although I wasn't ready to attack the summit quite yet, a viable route to the top seemed to be emerging.

The proof, as I structured it, rested on four separate estimates to the critical complex Monge-Ampère equation: the so-called zeroth-order, first-order, second-order, and third-order estimates. The solution to a Monge-Ampère equation, as I've said, is a function, and the whole point of this exercise is to establish some bounds on the function that show it cannot get too big (in the positive direction) or too small (in the negative direction)—to show, in other words, that the function cannot go to infinity. The zeroth-order estimate tells you the maximum value the function can attain. The first-order estimate tells you the maximum value of the function's first derivative. One needs to demonstrate, more specifically, that the first derivative cannot get too big, which is equivalent to showing that the function cannot fluctuate too rapidly. The second-order estimate, similarly, tells you the maximum value of the function's second derivative. One must show, again, that this estimate is bounded, meaning that the first-order derivative does not fluctuate too wildly. The same idea holds for the third-order estimate and up. These higher order estimates provide information on how the function changes—how big those changes are and how rapidly they occur.

In the summer of 1975, just before going to New York, I had succeeded in pinning down the second-order estimate. During the few months

I spent at Courant, I made something of a conceptual breakthrough. I realized that, at this point, all I needed to get was the zeroth-order estimate because with the zeroth and second in hand, I could also derive the first-order and third-order estimates. In other words, the entire proof now rested on determining a single estimate, the zeroth order. And that estimate depended on my showing that the function could not get too big—that its maximum value would never exceed some constant value. Put in those terms, the solution to this immensely complicated conjecture, which only a select number of mathematicians understood at the time, seemed like a straightforward enough proposition. But getting that estimate, which would impose a literal and metaphorical ceiling on the function, was not as easy as it sounded.

I was unable to overcome this final obstacle while in New York, but Cheng and I did have success in another area: We found a solution to the higher-dimensional Minkowski equation, the same problem we had made some inroads on earlier in the year. Moser was excited that we had finished this work while at Courant, just as everyone at Courant, it seemed, was enthusiastic about any work of significance that was accomplished there. Moser asked Cheng and me to present our solution at a seminar, which came off well.

I subsequently learned that the Soviet geometer Aleksei Pogorelov had independently solved this problem by a completely different method. His paper came out before ours, but it was printed only in Russian, in a journal that is not widely known, so we hadn't heard about it. Although our paper was not first, it was not superfluous either because the method we developed was important in itself, apart from the result we had achieved, as it was later used to solve other problems in math.

And going beyond math for a moment (which is usually about as long as I can go), my three to four months in New York were extremely pleasant. I struck up a bit of a friendship at Courant with Eric Bedford, who had just received his PhD in math from Michigan. He showed me how to use the subway system. While we roamed the city, we talked about complex Monge-Ampère equations, which he was working on too, although his methodology was different from the more geometric approach I was following.

I enjoyed my walk every day from my apartment in SoHo through Greenwich Village to Courant. There were always interesting and unexpected things to see. For instance, I went past the same car parked on Spring Street over the course of several days. At first, the vehicle was perfectly intact. But a day later, the tires had been stolen. Over the next couple of days, the body of the car was stripped down more and more. Finally, the rest of the car was gone, only to be replaced by another vehicle that appeared to be in pristine condition, although it was anyone's guess as to how long that would last.

Little Italy was close to the apartment, and I enjoyed seeing the many festivals that were held there. I spent a lot of my free time with Cheng, his wife, and their baby boy, Bing. When we strolled together through Chinatown and other parts of the city, I often carried Bing (who later became my graduate student at Harvard, earning his PhD in mathematics in 2004). It was nice being so close to Chinatown, not only for the myriad dining possibilities, but also because I liked browsing in the bookstores. And on weekends I went to Princeton to visit Yu-Yun. All in all, my stay in New York proved to be quite convivial.

But in late December, I had to return to California. I flew to Los Angeles with Yu-Yun, who was interviewing with TRW, an aerospace company that has since been subsumed by Northrop Grumman. The interview went well, and the company soon offered her a position.

Yu-Yun went back to Princeton afterwards, and I returned to Stanford where my main research preoccupation, not surprisingly, was still the Calabi conjecture. I sensed that I was close to cracking the problem. The summit was in sight; it was just a matter of getting over that one last obstacle. I felt that if I kept pushing, I would eventually find a way to overcome it.

However, another thing was weighing heavily on my mind. In May 1976, after the spring term at Stanford ended, I visited Yu-Yun at Princeton with a specific goal: I sought, and requested, her hand in marriage—five and a half years after she had made an indelible impression on me in the Berkeley math library. It had been a long haul, and the two of us certainly had our ups and downs. But I'm happy to report that she said yes. We were officially engaged. My brother Stephen came to Princeton from Stony Brook to join us for dinner and celebrate the good news.

Not only had Yu-Yun accepted my proposal, she accepted TRW's offer as well, which would entail a move to Los Angeles in the near future, given that her job was starting in the fall of 1976. In the hopes of lining up something nearby, I contacted a friend at UCLA, the differential geometer Robert Greene, and told him I'd like to spend the year there. My Sloan Fellowship would cover my salary for the fall quarter, but I was hoping that UCLA would cover the winter and spring quarters if I taught classes. Greene told me that could easily be managed, and in this way I had laid the initial groundwork for our future: Yu-Yun and I would be able to live together and work in the Los Angeles area. She was impressed that I'd been able to arrange this so quickly, given how hard it was to get teaching jobs in those days. And I'm still grateful to Greene for helping to set me up in what proved to be a very hospitable work environment.

I stayed in Princeton until it was time for Yu-Yun to move in early July. We then packed her stuff and set off on a cross-country drive, accompanied by her mother and father. Our first stop was Washington, D.C., where we watched the Fourth of July fireworks during the nation's two-hundredth (bicentennial) birthday celebration. Along with a million other folks—a sizable fraction of whom appeared intoxicated, boisterously exuding patriotic fervor—we witnessed the pyrotechnics set off over the National Mall. Views of the Washington Monument and Capitol, as a backdrop to the flashing multicolored blazes in the sky, added to the splendor of the scene.

We drove to Boston next to see Yu-Yun's cousin, whose husband had just died. It was my first visit to that city, and I really liked it. I had no inkling that before long it would become our home for what's now been more than three decades.

We made another stop in Ithaca, New York, to see another of Yu-Yun's cousins. From there, we began our cross-country trek in earnest, making it a big sightseeing trip for her parents. We went to Yellowstone National Park and then followed the Rocky Mountains south to the Grand Canyon. Afterwards, we picked up Interstate 40 in Flagstaff, Arizona, and took it all the way to Barstow, California, where we switched to Interstate 15 en route to Los Angeles. The views along the way were stunning, accentuated by the fact that I was in love, eagerly anticipating the married life to come.

Throughout most of the trip, however, my mind quietly wandered toward mathematics. When I was driving, I thought in particular about a classic problem in topology, the Poincaré conjecture, and the fact that no one had yet figured out a good way of approaching it. The original formulation of Poincaré's problem, which concerns the precise definition of a sphere in topology, had not been solved at that time. The conjecture specifically posits that a "compact," three-dimensional surface (or manifold) —one that is bounded and finite in extent—is topologically equivalent to a sphere if every loop that can be drawn on that surface can be shrunk down to a point without either tearing the loop or tearing the surface. We call such a surface "simply connected," which is another way of saying that, unlike a donut, it does not have one or more holes. Using that terminology, the conjecture can be restated as follows: Is any compact, simply connected, three-dimensional surface the same, topologically speaking, as a sphere? While the problem may not sound that intimidating, little progress had been made on this question since it was first raised in 1904.

One might suppose that I would have focused on the Calabi conjecture instead, which was then my main preoccupation, as it had been for many years. I had given that problem much more attention, in part because it is more general, and I sensed it could lead to a large class of manifolds that we didn't know about. But I always like to have several problems to think about; if I get stuck on one problem, I can turn to something else. And if the problems are of a similar nature, sometimes an idea I get while thinking about one can be applied to the other.

Furthermore, I knew that the zeroth-order estimate of the Calabi conjecture, which I then saw as the lynchpin to the entire problem, would require elaborate calculations with paper and pencil—something I could not attend to, safely or competently, with my hands on the steering wheel. That's why I picked a more conceptual problem with which to engage the mathematical portion of my brain, and Poincaré's conundrum served that purpose well. A concrete approach for tackling the problem had yet to be worked out, and perhaps the best thing to do at that stage was to dream about it, which I did—while also trying to keep at least a portion of my mind on the road.

Altogether, we drove more than four thousand miles on our roundabout journey from Princeton to Southern California. And during much of that excursion, my thoughts ineluctably turned to Poincaré's problem (which I will discuss in some depth in Chapter 11). I didn't have a great breakthrough, I'm sorry to report, but I was correct in my supposition that geometric analysis could eventually offer a way in.

After arriving in Los Angeles in mid-July, we rented a three-bedroom apartment in Long Beach while we looked for a house. There wasn't a lot of time, as our wedding was set for early September, and we wanted to be settled in before the big day. We soon found a place in Sepulveda, a former agricultural area in the San Fernando Valley, which was rather far from the ocean. It was a bit of a drive to UCLA too; one could make it in about a half hour, without traffic, but the words "without traffic" and "Los Angeles" can rarely be juxtaposed in the same sentence. It could often take an hour or more, and Yu-Yun had an even longer drive to TRW, which was located in Redondo Beach. It would have been nice to have found something more conveniently situated, but that house—the first I ever purchased—was the only thing we came across that was remotely affordable while also offering some of the amenities we'd been looking for.

We then had a mad scramble, not much more than a month to get the house in order and prepare for the wedding. I drove all over the place looking for used furniture and other essential items, while Yu-Yun attended to her wedding gown and various wedding-related matters. Her parents stayed with us, as did my mother and brother Stephen, both of whom arrived about ten days before the wedding. My mother flew in from Hong Kong, and Stephen came from Harvard, where he had just become a Benjamin Peirce instructor, after having received his PhD in mathematics from Stony Brook a few months before.

The wedding took place on September 4, 1976, followed by a lunch for family and friends. I told Chern I was getting married, thinking he would not come because it was going to be a very modest affair, but he and his wife showed up, which pleased me. My friends Robert Greene and Bruce Bennett came too, as did my mother's cousin and her husband, who lived in California.

Yu-Yun and I had arranged to go to Catalina Island for our honeymoon, but we had to cancel at the last minute because we underestimated L.A. traffic and missed the ferry. So we went to San Diego instead, where we had a very nice time though also a very brief one, because two days later we had to return to work.

I was happy to get back to my research, which is usually the case, but this time perhaps more than ever because of the commotion in our house with Yu-Yun and me, her parents, and my mother all living under the same roof. I holed up in my study for as long as I could, pouring all of my energy into the Calabi conjecture. Within a week or two, the zeroth-order estimate had been completed and, consequently, the problem as a whole had been completed too. I was relieved and happy, as well as somewhat surprised, because the last few steps had fallen into place more quickly than I had expected.

People have asked me what it felt like to prove the conjecture after having thought about it, off and on, for more than six years. For some strange reason—perhaps influenced by the spirit of my father—my mind turned to an essay written by the Chinese scholar Wang Guowei, who had died about fifty years earlier. Guowei drew on excerpts of classic Chinese poems from the Song Dynasty (960 to 1279) to chart the three stages one typically goes through to achieve success in a major pursuit: First, the narrator scales a high tower, surveying the land in all directions, as far as he can see. He then notes how weak and thin he has become during his lonely quest, though feeling certain that the prize he seeks is well worth the sacrifice. Finally, while searching through a crowd, a thousand times or more, he catches a glimpse of "her"—the object of his pursuit—in the dim and fading light.

These passages summed up rather succinctly—and poetically—the stages I went through in proving the Calabi conjecture. I first needed a good vantage point from which to gain perspective on the problem as a whole. I worked hard—at times to the point of exhaustion, going for long stretches without sufficient food or rest—in pursuit of the quarry at hand. And later, in a fleeting moment of insight, I was able to see my way to the end.

Owing perhaps to my recollection of Guowei's essay, I also began to think about another famous poem from the Song Dynasty that captured my sentiments after the conclusion of the Calabi proof. The poem depicts a scene in a garden in late spring, sometime long ago: As flower petals drop delicately to the ground, two swallows hover above, flying together in unison. That image resonated with me because solving this mathematics problem had curiously given me a new understanding of and appreciation for nature. By virtue of this work, I felt at one with nature—a sense that was conveyed in the image of two swallows flying as one.

That's sort of what I was experiencing at an emotional level, but on an intellectual level I wasn't yet ready to call this a triumph. I had been burned once before with the Calabi conjecture, having assumed three years earlier that I had proved it wrong, only to learn that I'd been mistaken. This time, I didn't want to take any chances. I checked and rechecked the proof in painstaking detail, going through it four times in four different ways, telling myself that if I got it wrong this time, I would give up mathematics altogether and try my hand at something different—maybe even duck farming. I also sought outside verification. I mailed a copy of the proof to Calabi and made arrangements to follow that up with a visit later in the fall to the University of Pennsylvania.

In the meantime, my UCLA colleague David Gieseker, whom I'd known at IAS, told me about a forthcoming talk in late September by the Harvard algebraic geometer David Mumford. It took me more than two hours to reach the University of California, Irvine, where the seminar was held, but I always think it's worth hearing what good mathematicians have to say. Mumford focused on a particular "inequality"—a mathematical expression in which one term was less than or greater than the other. The original inequality had been posed about a decade earlier by Antonius van de Ven of Leiden University, but Mumford also mentioned the contributions to this problem made recently by the Russian mathematician Fedor Bogomolov.

At some point during these remarks, I realized I had come across this inequality before—in my early attempts to *disprove* the Calabi conjecture—and I was pretty sure it could be stated in the exact terms Mumford had specified. I spoke with him after the seminar, telling him I thought I had proved the very point he had raised. I'm pretty sure he didn't believe me, as I was young and unknown in the world of algebraic geometry. But when I got home, I reviewed my calculations and found that I had used this same inequality during one of my attempts to find a counterexample to the Calabi conjecture. Now that the conjecture had been shown to be true, its corollary—which I had once attempted to disprove in the hopes of establishing a counterexample—must be true as well. That meant I had indeed proved the formulation Mumford spoke of, which is sometimes referred to as the Bogomolov-Miyaoka-Yau inequality. The question as to what happens in the special case when this inequality actually becomes an equality was an open problem, but my method of proof provided a complete determination of the circumstances under which that could occur. This determination in turn led me to the solution of a well-known problem, dating back to the early 1930s, called the Severi conjecture.

I sent a letter to Mumford the next day, laying out my argument. He showed it to his Harvard colleague Phillip Griffiths, and they both agreed that my reasoning was sound. News of these findings spread quickly. At first, people were much more excited about the proofs of the inequality and Severi conjecture than they were about the Calabi proof, even though I insisted that the Calabi conjecture was far more important.

Robert Greene, whose UCLA office was next to mine, appreciated the significance of the latter effort—as did most of the mathematics community in time—and Greene was especially happy I had achieved these results while working at his university. Some algebraic geometers were not thrilled because, in the course of solving two high-profile problems in algebraic geometry, I had not used any of the standard methods of their field. Mumford was different in this regard because he had an open mind, and I believe that's part of the reason Harvard offered me a job two years later.

This work instantly made me famous, or at least raised my profile, within the math community, and various offers and opportunities started coming my way. The mathematician Isadore Singer contacted me around that time to see whether I might spend a month or so at MIT, starting in November. As I still had the Sloan Fellowship, I was free of teaching duties, so I decided to take Singer up on his offer.

Before going to MIT, I stopped off in Philadelphia to see Calabi and his colleagues, walking them through the proof, step by step. Jerry Kazdan, a mathematician on the Pennsylvania faculty, took detailed notes during my presentation and, to my dismay, shared them with the French mathematician Thierry Aubin without letting me know. Aubin had independently proved a special case of the Calabi conjecture, but with the benefit of the notes provided by Kazdan, he went ahead and claimed credit for a proof of the full conjecture. Kazdan was later kind enough to set the record straight, stating in a published note that he had "learned of Yau's work during a lecture in December 1976" and subsequently extended that result in a joint paper with Aubin. In this way, Kazdan averted what might otherwise have evolved (or devolved) into an unsavory dispute.

At the conclusion of my meeting with Calabi, he said that everything looked good with respect to my proof. He was, and still is, a superb geometer but didn't have much expertise when it came to partial differential equations, so he felt it would be worthwhile for us to get together with Nirenberg too. The only time all three of us were free of other work commitments was on Christmas Day, so we agreed to convene in New York City at that time. Calabi claims that was the only time in his life he had a professional obligation on that day, although he happens to be Jewish, as is Nirenberg. I had never celebrated the holiday before, which may be one reason why the three of us were able to arrange a daylong work meeting on Christmas.

After my brief sojourn in Philadelphia, I made my way to Boston, stopping first at Yale in New Haven. Singer, who had invited me to MIT, was quite busy then, having to leave town on personal business for much of the time I was there. As a result, I saw only him once, for dinner, which turned out to be rather consequential. He was working with Michael Atiyah and Nigel Hitchin (my former IAS friend) on special solutions to the equations of C. N. Yang and Robert Mills, which were of fundamental importance to particle physics. Singer felt strongly about the unification of physics and mathematics, and he got me interested in this too. Several years later, in fact, I started working on solutions to the Yang-Mills equations as well. Some papers I did on the subject with Karen Uhlenbeck are considered rather important, and I have to thank Singer for pointing me in this direction.

However, other than that one meal I shared with Singer during my stay at MIT, I was pretty much on my own, as there weren't too many other geometers around at that time. I was put up in a studio apartment, within walking distance of the school, and spent most of my hours writing up the full Calabi proof, while the snow piled up, rather beautifully, outside my window. Upon finishing this paper, I planned to send it to the Courant publication *Communications on Pure and Applied Mathematics*, out of gratitude to Moser, Nirenberg, and other mathematicians at Courant who'd been so courteous to me. I'd already finished writing a brief announcement of the proof, minus the technical details, which came out in 1977 in the *Proceedings of the National Academy of Sciences*. The prospects for that manuscript had been enhanced, no doubt, by the fact that it was originally sent to the journal by Chern, an esteemed member of the National Academy.

Harvard, just a mile and a half down the road from MIT, asked me to deliver a series of lectures on the Calabi proof. People at Harvard—including Mumford, Griffiths, and Heisuke Hironaka, as well as the visiting mathematician Andrey Todorov—seemed more curious about the Calabi conjecture than the people I'd met at MIT (except for Singer, who was preoccupied with other pressing matters). So I ended up spending more time at Harvard, and the university even put me up for another month after my month at MIT ended.

I still recall an interesting conversation I had with the algebraic geometer Hironaka (although this occurred at a somewhat later date) about pursuing mathematics in the United States as a person of Asian descent. "It is much easier for Asian Americans to get tenure at a good university in America than at a second-rate university," said the Japanese-born Hironaka, who won a Fields Medal in 1970 while on the faculty of Harvard. "Because at a second-rate university, where research is not a priority, job advancement is based more on other things like golf." As someone who has never picked up a golf club in my life, I drew some comfort from those words. For if I truly excelled in my work, I might not have to take up that sport, which was probably never going to be my strong suit.

All told, I enjoyed myself at Harvard and particularly liked the collegiality within the math department. Before I knew it, Christmas was approaching, and I headed to New York for the meeting with Calabi and Nirenberg and my rendezvous with destiny. The snow was coming down hard, and we spent the entire day going through the proof in Nirenberg's office, taking a break for lunch in Chinatown, which was about the only place we could find restaurants that were open. By the end of the day, my argument was still holding up; no flaws had been uncovered. Calabi and Nirenberg said they would review the manuscript further, but neither they nor anyone else since has found any problems. I published the abbreviated version of the proof in 1977, as mentioned, and the extended version a year later. The argument still stands.

Calabi claimed that the conclusion of our daylong meeting in New York, which yielded a firm consensus that the proof was valid, was the best Christmas present he'd ever received. I could certainly say the same. And I felt that 1976 was ending on a very good note, except for one thing: I missed Yu-Yun badly, after two months away from her. It was time for me to head back to Los Angeles and tend to my marriage under the hopefully forgiving warmth of the Southern California sun.

It might be worth reflecting for a moment on what the proof of the Calabi conjecture accomplished. It showed, for one thing, that nonlinear partial differential equations and geometry could be combined to good effect—a premise that had been driving my research for quite a few years. I also proved the existence, mathematically speaking, of a large class of multidimensional spaces postulated by Calabi—spaces endowed with a combination of distinctive properties that until then had not been considered possible. At the same time, the proof offered not just a solution to the Einstein equations in the case when matter is not present, but the largest class of solutions to those equations that we know of.

Ever since Einstein invented general relativity in 1915, according to the physicist and computer scientist Andrew Hanson, "we have struggled to find manifolds, or 'Einstein spaces,' that satisfied his demanding equations. For years, it was hard to find *any* solutions, yet here, remarkably, was a simple prescription for finding them in any dimension—a long, and possibly infinite, list of manifolds that are absolutely guaranteed to solve Einstein's equations."

Sometimes the proof of a theorem marks the end of a chapter. This happened in 1952 when "Hilbert's Fifth Problem," which was posed in 1900 by the great mathematician David Hilbert, was finally solved, thanks in large part to the efforts of the Harvard mathematician Andrew Gleason. The solution in that case required feats of great ingenuity, but rather than inspiring new research, it killed off much of the work in that sector of mathematics, leaving other investigators with little to do in the way of follow-through.

From early on, I felt that the Calabi conjecture was different because it would tap into a realm of geometry that was both deep and expansive. The solution to this problem, accordingly, would forge openings into other areas of mathematics that were ripe for exploration. This wasn't just wishful thinking but rather a consequence, in part, of the unusual way in which I had approached the problem. As you may recall, I first tried to prove the conjecture wrong by providing counterexamples. Given that the conjecture was correct, and had been proved so, all of those attempted counterexamples must, logically speaking, be correct too. In other words, they were theorems in their own right, and my initial announcement of the Calabi proof also unveiled the proofs of five associated theorems in the field of algebraic geometry. The most important of these, as discussed before, was the proof of the Severi conjecture, which had been unsolved for more than four decades. In addition, about a half dozen other problems in algebraic geometry—admittedly of lesser significance—were instantly solved as well. The upshot of this was that a conjecture once considered "too good to be true," had turned out to be even better than initially supposed.

Yet even that was not the whole story, because deep down, I had a vague though gnawing sense that the Calabi conjecture and its proof would conjoin with physics in an important way, apart from the link with Einstein's general relativity that I already knew about. I had no clue as to what form this connection might assume, though I still felt certain it was there—or somewhere to be found. Well, it took about eight years for physicists to establish the sort of tie-in with the "Calabi-Yau theorem" that I dreamed of, but it turned out to be well worth the wait.

CHAPTER SIX

The Road to Jiaoling

WHEN I WAS YOUNG, my favorite book was *Dream of the Red Chamber*. I'm probably not alone in that opinion, as this work is widely regarded as the greatest novel in all of Chinese literature. Written in the 1700s by Cao Xueqin (and possibly finished by others after Xueqin's death in 1763), the *Red Chamber* charts the rise and fall of the Jia family, whose dwindling fortunes paralleled the overall decline of the Qing Dynasty. It's a big, sprawling work—120 chapters spread over five volumes and a couple of thousand pages—involving a complicated array of intersecting plot lines. I started reading the novel when I was ten and was mesmerized by its depiction of life and society in eighteenth-century China.

I was moved by the love story at the center of this saga, while also relating to the class struggles it depicted, as my family constantly fought to maintain its lofty values even as our economic status waned. What I didn't realize back then was that the structure of this novel would influence the way I approach mathematics. The story contains hundreds of different threads and introduces hundreds of different characters. It takes a while, and some measure of discernment, to see how these distinct strands and individuals relate to each other, combining to produce a complex and multifaceted though fully integrated whole.

I see mathematics, particularly my efforts in geometric analysis, in a similar light. By this point, in 1977, I had proved several theorems, and over time I would prove several more. Most appeared to be independent from one another, but I saw a unifying structure in geometric analysis that established a link between these separate theorems. The same could be said for mathematics itself. The field has different branches that might seem unconnected until you step back far enough to realize they are all part of the same vast tree—not unlike the family tree that could be drawn to trace the lineages of the Jia clan from the *Red Chamber*. I was striving to have a general familiarity with the entire "tree" of mathematics, while focusing my attention on the newly budding branch of geometric analysis, which was an extension of the longer and broader limb of differential geometry.

On that note, it's fair to say that I have not yet mentioned what I consider to be the biggest accomplishment of the Calabi conjecture proof and of the related theorems that were proved almost simultaneously. Collectively, they constituted the first major successes of geometric analysis and thereby demonstrated the potential of this new method.

In the 1950s, the Japanese mathematician Kunihiko Kodaira helped forge a method for solving problems in geometry through the use of linear differential equations. Kodaira was building on the prior work of people like Hermann Weyl and William Hodge. Many others, including Michael Atiyah and Isadore Singer, subsequently made key contributions as well. I, meanwhile, was advocating the use of *nonlinear* differential equations, raising the possibility that geometric problems that could not be attacked by linear methods might now be solved.

My initial success in this area helped elevate the stature of geometric analysis as a whole, encouraging other researchers to give geometric analysis a try or at least take it seriously. I started working with a group of friends, and after we obtained some significant results, the idea took off from there.

One collaboration on this front, which worked out quite well, came about through a random encounter in UCLA's math department, which I returned to after my 1976 Christmas Day meeting with Calabi and Nirenberg in New York. I unexpectedly ran into Bill Meeks, with whom I'd been friendly at Berkeley. After chatting a bit, it became apparent that we shared an interest in minimal surfaces, and we made a plan to work together on the subject.

But I first checked out the course Meeks was teaching on three-

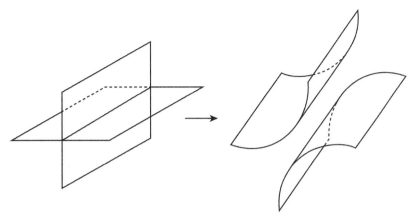

Dehn's lemma, versions of which were proved by the author and William Meeks, provides a mathematical technique for simplifying, or disentangling, a surface that may cross itself, converting it into a surface with no crossings, folds, or other "singularities." (Based on original drawings by Xianfeng [David] Gu and Xiaotian [Tim] Yin.)

dimensional manifolds. In the class I attended, he talked about Dehn's lemma, which I was already interested in. The German mathematician Max Dehn proposed in the early 1900s that if a disk had a "singularity"—a surface that crosses itself in a wrinkle or fold or another imperfection of that sort—it could be replaced by a disk that shared the same boundary circle yet had no singularity. Dehn's lemma was proved in 1956 by the Greek mathematician Christos Papakyriakopoulos (who was then based in Princeton)—a feat lionized in a limerick penned by John Milnor:

> The perfidious lemma of Dehn
> drove many a good man insane
>   but Christos Pap-
>   akyriakop-
> oulos proved it without any pain.

Meeks and I found a way to bolster this lemma. We used techniques developed in proving it to show that a large class of minimal surfaces has no singularities—building upon the prior work of Jesse Douglas, corecipient of the Fields Medal in 1936, the first year the medal was awarded.

Our strengthened version of Dehn's lemma was also a key piece in the eventual proof of a long-standing problem in topology, the Smith conjecture, which had been unsolved for almost four decades. First raised in 1939 by the American topologist Paul Smith, the conjecture concerns rotations of a three-dimensional space such as a sphere, not around a straight-line axis (as displayed in the world globes one might find in living rooms and libraries) but rather around a *knotted* line. Smith's statement—that such rotations were impossible—seemed intuitively obvious, for how could you rotate a sphere around a knotted axis? But it took our results, combined with the results of other mathematicians, including Cameron Gordon and Bill Thurston, to prove that the conjecture was actually true in a three-dimensional setting. As far as I know, this was the first time that minimal surface arguments had been successfully applied to a problem in topology. That provided encouragement for me and others to seek additional applications of this approach.

I enjoyed working with Meeks in part because he derived so much pleasure from mathematics. Many people in the field didn't give him the respect or recognition he deserved, despite his first-class work, because of his freewheeling lifestyle, which led them to believe he was not a serious scholar. That could explain why Berkeley refused to consider him for a permanent job, despite my recommendation, even when the department was specifically looking to hire a geometer. But Meeks didn't give a damn about what others thought, and his confidence never wavered. "I feel that if I really want to solve a problem in mathematics," he once told me, "I will solve it. And so far that attitude has never failed me."

We continued our collaboration at UCLA during the first half of 1977 until, after fulfilling his teaching obligations, Meeks left to take a visiting professorship in Rio de Janeiro. He also ended up exploring some business (and romantic) possibilities in Brazil. While his business venture did not work out, he fared better on the romance front. He became involved with two women in Brazil, and married one of them.

But Meeks also loved mathematics, and I'd seen that passion in other Americans too, who pursue the subject for the pure joy of it and can't imagine doing anything else. I know plenty of Chinese people who view mathematics as a good career choice—more of a means to an end for them than an end unto itself. They rarely get excited about the field in the same way; for them, it's a job rather than a passion.

Meeks, however, shared my enthusiasm for geometric analysis. He called our early investment in this discipline "a big gamble" that had paid off far better than could reasonably have been expected. And it's true that, thanks to the Calabi conjecture proof and other recent publications, I started to receive a fair number of job offers. Salomon Bochner, Calabi's former teacher, tried to lure me to Rice University, but I wasn't eager to move to Houston. Wu-Chung Hsiang, as I mentioned, called to say that Princeton wanted to hire me—a curious turn of events, given my recent encounter with him over office space (or the lack thereof). It was a nice offer, nevertheless, even though I ended up declining it, in part because my wife was happily employed on the West Coast.

UCLA also wanted to hire me, although that offer was complicated by the fact that I had received terrible student evaluations in the first course I taught there. I had a large group of economics and humanities majors in my class who had no interest in mathematics and felt they could chat whenever they liked. I warned them that I might give an exam at any moment, without notice. I never made good on that threat, but it got their attention—and also made me very unpopular among this group of students. By the end of the course, they had learned the material surprisingly well. (The prospect of a hanging, as Samuel Johnson once said, "focuses the mind wonderfully.") But these students never liked me, and their evaluations were consequently quite negative. Some graduate students from Stanford had to let the UCLA math department know that I could actually teach.

I was still a tenured faculty member at Stanford and figured that if I were to teach anywhere in the University of California system, it would probably be at Berkeley, my former home. Both Singer, who was then visiting Berkeley from MIT, and Chern came to Los Angeles to try to lure me up north. Chern was flexing his muscles, as he was prepared to offer me a position at "Step 6," which was quite high up the academic ladder for someone in his mid- to late twenties. It normally takes a number of strong letters of recommendation before a professor gets promoted to that rank. Other people in Berkeley's math department, some of whom had taught there for many years without attaining the Step 6 designation, were unhappy that I, as a relative newcomer, had received such an offer.

Yu-Yun still had her job with TRW in Los Angeles and didn't have anything lined up in the Bay Area, so I decided not to make a major career shift at that moment, instead electing to retain my position at Stanford and go to Berkeley for one year as a visiting professor. My mother stayed with me in Berkeley during the 1977–1978 academic year, while Yu-Yun remained in Los Angeles with her parents. S. Y. Cheng used his Sloan Fellowship to come to Berkeley that year, and Rick Schoen, who'd just received his PhD from Stanford, came to Berkeley as well, as an instructor. So I was lucky to have some of my closest collaborators nearby.

I later started working with Peter Li, one of Chern's graduate students, who came from a wealthy family in Hong Kong. Li had a fancy car, an Alfa Romeo, and Chern had asked him to drive me around, as needed, to keep me happy. And even Meeks came up from Brazil, planning to work with me for a couple of weeks. He stopped by my apartment regularly for meals, as did Schoen. My mother took care of the cooking, which was good, as my proficiency in that area was still wanting.

One night I had some big shots over for a dinner party, including Stephen Smale of Berkeley, who'd won the Fields Medal a decade earlier for his proof of the higher-dimensional Poincaré conjecture, and Singer, who was still visiting the department. Schoen came too, as did Meeks, who showed up with a barefoot woman he had just met. Meeks was perfectly relaxed about bringing an unannounced guest, not embarrassed in the slightest. I think that shows how different people in California were from those on the East Coast, as this sort of thing probably would never have occurred in the somewhat stuffier, more patrician climate at Harvard. My mother was often confused by the carryings-on of my guests, but she took it in stride, never letting it affect the quality of her cuisine.

One evening, late in the fall of 1977, Schoen and I headed from my Berkeley office to the apartment for dinner, and along the way we had an idea about the positive mass conjecture—a problem I had been introduced to during the aforementioned talk by Robert Geroch at the 1973 Stanford conference. The conjecture held that the total mass or energy must be positive in any isolated physical system, even the universe itself. Many physicists, including Geroch, believed the statement had to be true, and he challenged geometers to find a proof to this long-standing problem in general relativity.

Some geometers, on the other hand, argued quite forcefully that the statement, in its full generality, could not be true. I was not willing to simply accept these skeptical pronouncements and leave it at that. To me, the proposition warranted further exploration, and I had some vague ideas on how to proceed.

General relativity, which concerns itself with the curvature at every point of space-time (or, loosely speaking, at every point of the universe), is a highly nonlinear theory, as noted before. What we wanted to prove boiled down to this: The average curvature at each point of space-time must be positive. Schoen and I felt we might be able to make progress with nonlinear tools from geometric analysis, specifically minimal surface techniques, which had never been applied to this problem before. The fact that those techniques had not been applied was not surprising because there was no obvious link between the positive mass conjecture and minimal surfaces. We just had a hunch that the latter could provide some useful analytic tools for tackling this problem.

After encountering some initial difficulties, we eventually hit upon a two-step strategy: In the first step we proved that if the average curvature of a space-time is positive everywhere, then the total mass is positive too. In the second step we constructed a space-time with positive average curvature that had the same mass as our universe. Putting the two parts together, it was clear that the total mass of our newly constructed space-time was positive, which meant that the mass of our universe must be positive as well.

This was the line of attack Schoen and I followed in the spring of 1978 to solve a special case of this conjecture, the so-called time-symmetric case, which is the problem Geroch had proposed. The approach we followed, proof by contradiction, was the same general strategy I had tried in my unsuccessful attempt to prove the Calabi conjecture wrong. We assumed, for starters, that the mass of a given isolated space is not positive. We then showed that one could construct an area-minimizing surface within the confines of that space that had a special kind of curvature—zero average curvature, in fact—which was simply not possible in a universe like ours where the matter density is nonnegative. And if such a surface could not exist in the universe we inhabit, our original premise must have been wrong, leading instead to the opposite conclusion: the mass of any isolated space, or physical system, must be positive. One could just as well say that the energy of any isolated space must be positive too, for energy and mass are equivalent in general relativity, and that's exactly what we showed.

Many physicists, however, believed we would not get beyond the time-symmetric case. Stanley Deser of Brandeis and Larry Smarr, who was then at Harvard, told me that we had not really proved the positive mass conjecture until we solved the general case too. That's what Schoen and I took on in the summer of 1978 while I was back at Stanford, having finished my one-year stay at Berkeley. We borrowed a nonlinear equation that a Korean physicist named Jung had been studying, noting its similarities to the minimal surface equation we'd been wrestling with. Using this equation, Schoen and I showed how the general case of the conjecture could be reduced to the special case we had already proved.

The implications of this, completing the more general proof, couldn't have been bigger. If the total energy of the universe were positive, it would have a lower limit and would always remain somewhere above zero. If, on the other hand, the universe's total energy were negative, there would be no lower bound. It could keep dropping and dropping indefinitely, with nothing to stop it. That, in turn, would make the universe unstable until it would eventually no longer be able to hold together—an off-putting proposition, to say the least. It would be overstating things to say that, through our proof, Schoen and I had saved the universe, but our work did provide some reassurance toward that end. And as the second major success of geometric analysis, our proof also provided reassurance that this might be a fruitful avenue in mathematics. Many of the tools we developed in the course of solving this problem, moreover, are still being used today, and some people may consider those tools to be as important as the solution itself.

Nevertheless, our argument, which appeared in published form in 1979, didn't win much early support among physicists, perhaps because the nonlinear calculations were difficult for them to follow, and the same was true for many mathematicians. The University of Maryland physicist Bei-Lok Hu—who went to my high school in Hong Kong and later led the Chairman Mao study group I joined while at IAS—was among the possibly sizable number of researchers who simply did not believe our proof. Hu had received his PhD under the direction of the physicist John Wheeler, one of the world's foremost experts on general relativity, and he asked me point-blank, "How could a mathematician possibly prove such a thing?" Yet the proof has held up all this time, going on four decades now, and our credibility was quickly enhanced when Stephen Hawking invited me to discuss it with him at Cambridge University in late August 1978.

I gladly accepted, planning to make several stops in Europe before going to Cambridge, as I'd also been invited to Paris, Rome, and Finland, where I'd be giving a talk at the International Congress of Mathematicians in Helsinki. Travel was difficult, however, because the British Consulate had recently taken my Hong Kong resident card, maintaining that I could not keep it now that I had a U.S. green card. In the process, I had become stateless. I was no longer a citizen of any country, although I was a legal resident of the United States. For this period of time, until I became a U.S. citizen in 1990, I was literally a man without a state, stuck between two countries and between two cultures. Traveling abroad, as a result, became a real nuisance. I had to use my "white card" to apply in advance to leave the United States, and if I did not follow the proper steps, I would not be allowed back in.

I could not secure a visa to get into Italy, so I had to cross Rome off my list this time, even though I paid an extra "fee" to the Italian consul, who said he'd take care of it (and didn't). The same thing happened the next time I was invited to Italy; I again paid the consul an extra fee, but still got no visa. On another occasion, Michael Atiyah invited me to give a talk in Wales before the London Mathematical Society. When I showed my white card at immigration control in London, they gave me a hard time. "What is the purpose of your visit in the United Kingdom?" I was asked. I told them I was there for tourism. "Where do you plan to go?" I was then asked. Wales, I replied. "Why are you going to Wales," the border agent persisted, "since it's clearly not a good place for tourism?" The questions finally ceased when I explained that I was going to Wales with my good friend Nigel Hitchin, an esteemed professor at Oxford. That did the trick this time, though traveling with a white card caused me no shortage of headaches.

For this trip, in August 1978, I was able gain entrance to France, Germany, Finland, and finally to England to see Hawking and his colleagues. My first stop was Paris, where I met with French mathematicians—Jean-Pierre Bourguignon, Nicolaas Kuiper, and many others—at the Institut des Hautes Études Scientifiques (IHES).

I also saw Blaine Lawson, who was then visiting IHES from Stony Brook. I briefed him on some recent work I had done with Schoen following up on the positive mass theorem. That theorem concerned manifolds of positive (scalar) curvature, and Schoen and I delved further into the structure of such manifolds. In particular, I told him about our procedure for generating a large number of geometrically similar, three-dimensional manifolds through surgical techniques, pioneered by Milnor and others, which are somewhat like organ transplants in humans. The basic idea is to remove some internal portion of the manifold, such as a sphere, and replace it with something else—or another embedded sphere in a different position and possibly of another dimension—while maintaining the manifold's positive scalar curvature. The latter point is important in general relativity, where scalar curvature is related to the density of matter. Because the matter density must be positive, as Schoen and I proved, the space we live in must have positive scalar curvature as well.

We had also shown that if you take two (three-dimensional) manifolds of positive scalar curvature—like two universes, for example—and link them together with a tunnel or bridge, the result will be a new three-dimensional manifold (or universe) that also has positive scalar curvature. I described this method to Lawson in great detail, explaining how it related to the more general process I had discussed with him earlier.

The paper Schoen and I wrote was published a year later, 1979, in a relatively obscure journal, *Manuscripta Mathematica*, but the surgical approach we outlined has become an important tool in the study of manifolds with positive scalar curvature. It's well known that once a surgical technique like that is identified, many topological consequences follow. Schoen and I did not explore them in our paper because we were more interested in the implications for the positive mass conjecture and general relativity at large.

Lawson, meanwhile, teamed up with Gromov—who was then visiting IHES from his base at Stony Brook—to discuss the topological consequences of this sort of surgery in a paper that appeared in the *Annals of Mathematics* shortly after ours came out.

Kuiper, a geometer who was then directing IHES, invited me to have lunch with him and Robert Connelly, a Cornell mathematician who had made an important discovery within the previous year. Connelly was working on a topic that had been raised in 1766 by the great Leonhard Euler, who said, "A closed spatial figure allows no changes, as long as it is not ripped apart." At issue here was the question of whether a closed surface sitting in three-dimensional space could be "flexible" or not. In other words, could such a surface be continuously deformed without changing its internal structure and hence without changing its geometry? If the answer were yes, such a space would be classified as flexible.

A simple example might help illustrate this concept. If one were to take a flat piece of paper and gradually roll it up until it formed a cylinder, the surface would change throughout that process. But the geometry of the paper would stay the same because that depends solely on the distance between points on the paper, and the shortest distance *along the surface* between two fixed points stays the same, regardless of whether the paper is perfectly flat or rolled up into a tube. So the paper in this example would qualify as flexible.

In 1813, the French mathematician Augustin-Louis Cauchy argued that the surface of a three-dimensional, convex polyhedron—a surface that bulges outward at every point (like a fully inflated soccer ball) and is composed of polygonal faces that meet in edges—must be "rigid" rather than flexible. However, a concave polyhedron (like a bashed-in, deflated soccer ball) could, in principle, be flexible.

In 1977, Connelly presented the first example of a truly flexible polyhedron, consisting of eighteen triangular faces that are themselves solid and unbending. But the edges of the polyhedron, where two triangles meet, are like hinges that can curve inward or outward. As with the piece of paper in the previous example, the shortest distance along the surface between two points of the polyhedron does not change, regardless of how the triangular faces are positioned. Connelly's polyhedron thus met the criterion of flexibility that had eluded mathematicians since Euler's famous statement, which was made more than two centuries before. Connelly, along with Idzhad Sabitov and Anke Walz, later proved that the volume of the polyhedron stays constant at all times, even as the surface is being deformed.

He brought a model of his polyhedron, which is called the Connelly sphere, with him to Paris in 1978. Kuiper was very interested in this object, and he and Pierre Deligne (who was then at IHES) later modified it to create another flexible polyhedron with eighteen faces.

During our visit, Kuiper invited Connelly and me, as well as the American mathematician Ken Ribet, to meet some artists in Paris. We took Connelly's model with us. We were astonished to see that these artists had also created flexible polyhedra, which they had incorporated into

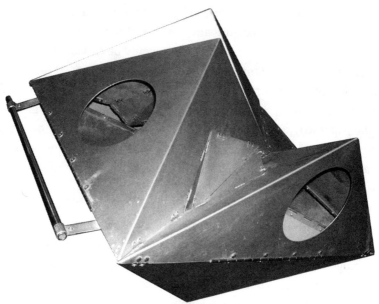

Euler conjectured in the 1760s that all polyhedra are rigid, but in the mid-1970s, Robert Connelly produced a counterexample—a polyhedron that could flex. Polyhedra of this type are inevitably concave and must have specific geometric features. This model of a "Connelly sphere," which was built at IHES, is based on Connelly's early work. Later, much simpler examples of flexible polyhedra were found. (This image, courtesy of Jean-Pierre Bourguignon and IHES, is from the forthcoming book *Frameworks, Tensegrities and Symmetry: Understanding Stable Structures*, by Robert Connelly and Simon Guest.)

sculptural pieces. Their work showed great insights into geometry, even though they were not formally schooled in the subject. While the motivations of the artists and mathematicians were almost entirely different, each group, in its own way, was engaged in the pursuit of beauty. And I suspect that the drive to produce something beautiful, or to uncover beauty in nature, is common to all humanity, regardless of one's profession or the country in which one lives.

That excursion into Paris was a revelation to me, and I tried to get into the city as often as I could, even though IHES was located about twenty miles away. One night, I was wandering around Paris with a Stanford graduate student who was also at IHES. We planned to see a movie called *Hitler* when we ran into the French mathematician Bernard Saint-Donat, who wanted us to accompany him to the opera. The student, however, insisted on seeing the movie, much to Saint-Donat's dismay over the two ill-bred Americans who would choose a movie (about a deranged, evil, and murderous tyrant) over the city's finer offerings in the performing arts. I can say, in my defense, that I later went to many outstanding museums in Paris and always try to do so when I'm in the city.

My next stop was Bonn, where I was asked to give a talk by Friedrich Hirzebruch, a great algebraic geometer. I was happy to get the chance to spend time with Hirzebruch, because I admired his work and had first learned about the theory of Chern classes from his book *Topological Methods in Algebraic Geometry*. While in Bonn, I also met Stefan Hildebrandt and Wilhelm Klingenberg, becoming friends with both of them. They later sent me some excellent students. I had the chance to visit other cities in Germany that were overflowing with the history of mathematics. I came away from my trip impressed by the rich math tradition in that country, which had been home to Carl Friedrich Gauss, Bernhard Riemann, David Hilbert, and other giants of the field.

During the train ride from Bonn to Frankfurt, where I was to catch a plane to Helsinki for the International Congress of Mathematicians (ICM), I sat next to the Japanese mathematician Tetsuji Shioda, who was also headed to the congress. We had a long time to talk and got onto the subject of Chinese characters. Shioda insisted that Chinese characters were useless, basing his claim in part on the fact that they could not be used with a typewriter. I argued the opposing case in our conversation, which sometimes became heated but always remained civil. When I met Shioda thirty-five years later in Tokyo, I was glad to hear that he had changed his mind and was willing to concede that Chinese characters had some value after all.

That trip in 1978 included my first visit to Finland, where I was scheduled to give a plenary address at the ICM. A number of prominent people were scheduled to give such talks too—including Lars Ahlfors, Robert Langlands, Roger Penrose, and André Weil—although, at age twenty-nine, I was probably the youngest member of that crowd.

A couple of days into the conference, I heard about something awful that happened at Stanford: A deranged graduate student, Theodore Streleski, had walked into the office of the mathematician Karel deLeeuw and killed him with a hammer. DeLeeuw was a nice man, a father of three, whose office was just two doors from mine. His murder was a horrible tragedy, and all of us at the congress were badly shaken by the news. The event continued as planned, of course, although a sense of sadness hung over the proceedings.

My talk was intended to be an introduction to the subject of geometric analysis—the gist of which was not widely known at the time. I planned to discuss the philosophy behind this approach and chart its development, underscoring the important role nonlinear differential equations could play in geometry. But when I found out how big the lecture hall was, I realized my preparations had not been adequate. I had assumed I could use a blackboard and deliver the kind of talk suitable for a classroom, but once I saw the size of the auditorium it became clear that a blackboard talk would be out of the question. I had gotten Stanford to hire my friend Yum-Tong Siu, a legendary student at my Hong Kong middle school and high school, and he was at the ICM to give a talk too. As a graduate student at Princeton, Siu had provided figures for John Milnor's book on singularities, and he was kind enough to draw a picture or two for my talk.

I also got some assistance from Bill Casselman, an American-born number theorist who settled in Canada decades ago. He loaned me a watch so that I could keep track of time during my lecture and not exceed the hour allotted to me. When I finished the presentation, Casselman immediately bounded onto the stage; I thought he was so excited by the talk that he couldn't wait to congratulate me or ask some questions. But he just wanted his watch back.

Years later, his most vivid memories of the conference had nothing to do with my lecture or with his own invited address ("Jacquet Modules for Real Reductive Groups") but instead concerned the more than six-foot-tall blond Finnish woman who served us breakfast every day.

Nevertheless, my remarks appeared to have made an impression on some of the people in attendance. After the ICM concluded, I flew from Helsinki to London for my meeting with Hawking. I sat next to Chern on the plane, who in turn sat next to Lipman Bers, a well-known mathematician at Columbia. Bers paid me a backhanded compliment, telling me my talk was "the second best at the whole conference." The best talk, in his opinion, was Thurston's "Geometry and Topology in Three Dimensions." I agreed with Bers on one point: The work discussed by my former classmate at Berkeley had indeed been very important. I felt the same about the work that I had presented, although I did not quibble with Bers over the ranking.

I was excited to speak with Hawking, who had become, by virtue of his work on black holes—especially his ideas concerning black hole (or "Hawking") radiation—one of the most famous scientists in the world. We got together my first morning in Cambridge, sitting in a garden outside his university suite. Hawking asked me a lot of questions about the positive mass conjecture, although a student had to translate for him because his speech—impaired by the disease he has long suffered from, amyotrophic lateral sclerosis (ALS)—was difficult to decipher.

But Hawking, who was then in his mid-thirties, was still extremely energetic. His mind was lightning quick, even though his movements were becoming increasingly limited because of the steady muscular degeneration that accompanied his condition. Knowing I wouldn't have an opportunity like this too often, I asked Hawking as many questions as he could comfortably entertain. He was a charming and witty host, as well as a brilliant scholar, and I'm grateful for our interactions. (I, along with the rest of the world, mourned Hawking's death in March 2018. He truly was an inspirational figure, who showed how much a person can accomplish, and how fully one can live, despite a crippling disability.)

Schoen and I initially proved the positive mass conjecture in three dimensions, but Hawking was particularly interested in the four-dimensional version of the conjecture since space-time in general relativity consists of four dimensions, three spatial and one temporal. Hawking and his physics colleague Gary Gibbons were developing a new gravitational theory called "Euclidean quantum gravity," and their scenario was predicated on the energy of four-dimensional space-time being positive. Hawking, accordingly, wanted to know whether an argument similar to what Schoen and I had originally cooked up would still hold in the higher-dimensional case.

I was not able to answer that question offhand but was hopeful that a slightly modified approach would work. When I got back to Stanford, I took up this problem with Schoen, and within a couple of months, we had proved the positive mass conjecture in four dimensions—a result I was happy to share with Hawking.

Schoen and I also started to follow up on some work of Hawking and Roger Penrose that dated back to the late 1960s and early 1970s. In a series of papers, Hawking and Penrose described the precise circumstances in general relativity that can produce a singularity—a place, such as at the center of a black hole, where gravity, curvature, and matter density all approach infinity. Hawking and Penrose proved, through a geometric argument, that a special kind of surface known as a "trapped surface" would lead to just such a singularity. A trapped surface is a collapsing surface whose "walls," so to speak, are rapidly closing in as its area goes to zero and its curvature goes to infinity.

Schoen and I took this a step further, trying to ascertain the conditions that would give rise to a trapped surface in the first place. We showed—once again using geometric arguments, though of a different type than those Hawking and Penrose used—that a trapped surface will automatically form in a region whose density is twice that of neutron stars, so named because they are composed almost entirely of neutrons. The smallest and densest stars known to exist in the universe, neutron stars are more than one hundred trillion times denser than water. (Put in other terms, a teaspoon of neutron star material would weigh more than a billion metric tons—about five hundred times heavier than the Great Pyramid of Giza.)

Our result, combined with the earlier findings of Hawking and Penrose, spells out the conditions under which a black hole would inexorably materialize. In other words, we had demonstrated through mathematics that black holes must exist, and we did this before these objects were experimentally confirmed. Astrophysicists now believe that black holes are extremely common, with giant ones lurking in the center of practically every large galaxy. Proving the existence of black holes is, in my opinion, an important contribution made through geometry toward understanding our universe.

While wrapping up this work, I lived in a small house I had purchased near Stanford. My mother came to stay with me there. Yu-Yun, meanwhile, had just moved to San Diego to take a job with a small company based in La Jolla called Physical Dynamics. We sold our house in Los Angeles and bought one in Del Mar, about twenty miles north of San Diego, where Yu-Yun and her parents lived. It might sound like an odd arrangement, living in separate homes with our respective parents, but it wasn't such an unusual situation for Chinese families to find themselves in.

I have failed to mention thus far that my departure from Berkeley had been anything but smooth. Chern wanted me to remain there permanently, which was flattering and very generous. He let me know that if I did the right thing, that is, stay at Berkeley, I would be his anointed successor.

At that time, Chern, Singer, and Calvin Moore had big plans to create a new math center at Berkeley, the Mathematical Sciences Research Institute (MSRI), partially funded by the National Science Foundation (NSF). But they had to overcome major resistance from the powers that be at IAS, who felt that if such a center were built anywhere with financial backing from the NSF, it ought to be located in the best and most prestigious setting—that being Princeton. IAS personnel lobbied forcefully, on multiple fronts, to make that case. It was an all-out brawl, with Saunders Mac Lane of the University of Chicago vying for a center as well, but Chern, Singer, and Calvin ultimately prevailed: The Berkeley-based MSRI officially opened its doors in 1982. Chern served as its first director, and he told me that if I stayed at Berkeley, I would in all likelihood follow him in that role.

But things didn't turn out that way, as the fit wasn't quite right for me—partly because I was still focusing my energies on mathematical research and had little appetite for administrative chores or the political wheeling and dealing involved in running, or helping to run, a major mathematics center. Furthermore, although Berkeley had a big math department with a lot of great people in it, not many people there shared my specific interest in nonlinear partial differential equations and geometry, especially now that Schoen had completed his lectureship there and was headed to Courant. I suggested that Berkeley recruit Leon Simon, who was then at the University of Minnesota, but Chern said he couldn't get him hired because the department was then focusing on different aspects of mathematics.

I politely told Chern in the spring of 1978 that I could not stay at Berkeley because I felt I did better when I had a group of people to work with, and the people at Berkeley were interested in other topics. My comfortable arrangement at Stanford made it easy for me to pursue my research and train my students. I did not think I would be happy at Berkeley, and my productivity would suffer as a result.

Chern yelled at me, the first time I ever heard him yell at anyone, telling me how privileged I had been to be under his wing. Without his support and protection, he told me, my status in the mathematics community would have been very different. If I stayed, on the other hand, I would take his place as a leader in the field and would do so with his blessing.

It was hard to say no, as I was grateful for everything that Chern had done for me, and make no mistake, he had done a great deal on my behalf. But I was more interested in pursuing my own research—gaining whatever influence I had through my work in the subject itself—rather than trying to lead others. And that, I believe, was the fundamental point of divergence between Chern and me. At that stage in his career, in his late sixties, Chern was trying to shape developments in the field from above, calling the shots from the upper echelons of power. I was in my late twenties and didn't care much about that, hoping instead to make my mark at the ground level—or perhaps I should say at the paper level, with a pencil (or typewriter) as my principal instrument of choice.

I told Chern three times that I would be leaving Berkeley, but he refused to believe it. I did not want to make him unhappy, but after going back and forth in my head over the course of several months, I decided to move on.

My problems with Chern started in earnest from that point forward, although I had the sense that some of the people around him had already tried to stir up trouble between us. At a dinner some months earlier, I recall Wu-Yi Hsiang asking Chern, in front of me and everyone else, about his recent trip to China. As I remember it, Hsiang asked Chern whether he had told people that—in light of my proof of the Calabi conjecture—I had surpassed him in mathematics. Chern was shocked and turned bright red. I was embarrassed, too, because Chern and others might have thought that I had encouraged this sort of talk. This was just one example of what struck me as repeated attempts by certain people to turn Chern against me—a campaign that ultimately prevailed.

Back at Stanford in the fall of 1978, I started to work with Yum-Tong Siu, who was newly installed there. Together, we solved an important problem in complex geometry, the Frankel conjecture. Our proof relied on partial differential equations whereas an independent proof of a more general form of that conjecture by Shigefumi Mori of Japan relied entirely on algebraic geometry techniques. Siu and I were getting along well at the time, although he was extremely competitive, a competitiveness that ultimately took a toll on our relationship.

Back in those days, I was on the go a lot. In late March and early April of 1979, Peter Li arranged for S. Y. Cheng, Schoen, and me to attend a conference in Hawaii. It won't come as much of a revelation to say that some academic conferences can be pretexts to spend time in a beautiful place. Although I was extremely interested in the subject ("Geometry of the Laplace Operator"), I also tried to make the best of my time in the fiftieth state. After four fun-filled days in Oahu for the conference, we went to the spectacular island of Kauai for sightseeing. In his spare time, Schoen mastered the fine art of dislodging coconuts from palm trees by throwing rocks at them. We then faced the challenge of opening the coconuts after they reached the ground. Our extensive knowledge of topology, I'm sorry to report, was not much help in penetrating their crusty surfaces. In this case we would have done better by exchanging our equations for machetes.

Our return trip was delayed by a United Airlines strike, and we stayed together in a rental apartment for a few extra days. Late one night, thieves tried to break in, but Cheng said that my loud snoring scared them away.

When the strike was settled, I flew from Honolulu to Boston, where I spoke at Harvard about the positive mass conjecture. I stayed at the home of the MIT differential geometer Richard Melrose. We celebrated our thirtieth birthdays together, although mine was still a couple of days off. I then flew to San Diego where I happily celebrated my actual birthday, April 4, with Yu-Yun.

A couple of big things were looming on the calendar. Armand Borel, the Swiss mathematician who'd been a professor at IAS since 1957, had asked me to organize a "special year" on geometric analysis at the Institute, running through the academic year from the fall of 1979 through the spring of 1980. This was my chance to bring together the key people needed to really kick-start this field. But it was not just a matter of assembling the right personnel. I also had to come up with a structure for the proceedings that would maximize the yield. Although running this year-long symposium represented a tremendous opportunity for me, it would take a good deal of planning because of all the logistics involved. My preparations, of course, had already begun.

But 1979 would prove to be a special year for another reason. China was just opening up to the outside world, and the eminent scholar Loo-Keng Hua—who was in the midst of a long-term feud with my former advisor Chern—had invited me to give a series of lectures at the Institute of Mathematics at the Chinese Academy of Sciences in Beijing, starting in late May. This would be a momentous occasion for me, as I had not been to China since I was an infant, thirty years before. But I was not alone in this homecoming journey, as I would be joining a large number of expatriates who would be returning to their native land after similarly long absences.

I had a couple of weeks to spend in China in August before heading to IAS for the yearlong workshop on geometric analysis. When I landed in Beijing, I was so excited that I bent down, right next to the airplane, and touched the ground. It was a powerful moment for me because China was a huge presence in my life even though I had no actual recollections of being there.

I gave several lectures at the Chinese Academy on geometric analysis and other topics, fitting in some sightseeing in and around Beijing when I had time. I tried to hit all the highlights—the Great Wall, the Forbidden City, the Summer Palace, and many other spots I had conjured up vague mental images of even though I had never seen them. My visit was emotionally charged, to be sure, but not entirely blissful. Most people in China were still poor and uneducated, and their lives were very difficult—facts I could not ignore despite my celebrity treatment.

My talks at the academy proceeded without incident, but I did have an unpleasant experience in Beijing. It started with a visit from a mathematician who had formerly studied with Wenjun Wu. Wu had gained some renown for something he developed in algebraic topology called the "Wu class." As a protégé of Chern, he had been fiercely aligned against Hua. The fight between Wu and Hua led to a split in the mathematics program at the Chinese Academy. Wu was then in the process of forming the Institute of Systems Science as a separate center of math research at the academy, wholly distinct from the Institute of Mathematics, of which Hua had been the founding director. It seemed like a curious choice to me, as Wu was a pure mathematician, a topologist, who had little knowledge of applied math, but that just shows how pronounced the fissure between Hua and him had been.

During his meeting with me, Wu's former student showed me a paper he'd been working on. I casually said it looked nice, although I didn't have time to read through it in any detail. Wu then sent a report to the vice premier of China claiming that I'd said his student had done important work and should therefore be given a national award. Some of Hua's colleagues were upset that Wu's protégé had been singled out for this award—aided, perhaps, by my support. They asked my friend and Stanford colleague Siu to talk with me. Siu advised me to write a letter to the vice premier to correct the misimpression made by my apparent endorsement. I was reluctant to get more deeply involved in the conflict but finally did, writing a letter stating that this work, in my opinion, was not deserving of a major award. I had to go through the proper channels to get this letter into the hands of such a high government official. The protégé was not happy about the latest turn of events, and we had some tense discussions on this subject a year later.

Before I went to China, one of Hua's former students named Qi-Keng Lu had asked me what I wanted to do while I was in the country. I wasn't sure at first and conferred with some friends. "Obviously, you should go to your father's home and visit the tomb of your ancestors," a Chinese-born colleague told me. So I told Lu that I wanted to visit the village in Jiaoling County where my father had been born and where his ancestors had lived some eight hundred years before. (My siblings and I, as best we could trace it, were the twenty-third generation of the Yau/Chiu family from Jiaoling.) That request, which seemed modest enough, was greeted with a variety of excuses for not allowing the trip. First I was told that the town in question was not on the map. Then I was told it no longer existed. Next Lu told me the central government informed him that we could not go there for defense reasons. It felt like I was getting the runaround, though I had no idea why. I just wanted to visit my ancestral home, as my friends had advised, and that seemed like a reasonable request.

After much delay, I was finally given permission to travel to Jiaoling, accompanied by a mathematics professor from the Academy of Sciences named Wang. We took a roundabout way of getting there, stopping first at Guilin, a popular resort in southern China. While in Guilin, we took a brief river cruise, and I reveled in the stunning scenery—most notably the karst landscape, which consists of oddly shaped rocky hills that jut out of the ground and are adorned with lush green vegetation growing on the rocks' steep sides.

Wang was an amiable enough traveling companion, but the arrangements were somewhat awkward. Since I was an "honored guest" of China, I got to stay in much nicer rooms than he did, and when we went out for meals, we always sat at different tables, and I was served much better food. This made me feel uncomfortable, even though I was being treated quite generously.

After Guilin, we flew to Canton where I met a cousin of my father whose husband was a professor at the local university. With the help of others, she and her husband threw a banquet in my honor. It was all snakes—snake soup, fried snake, and so forth—which the Cantonese know how to cook. That was the first snake dinner I'd ever had, and after I got over my initial squeamishness, I found it quite tasty.

Afterwards, my hosts made a request that soon became a familiar refrain: They asked me to help their son go to college in the United States. I hesitated at first, because I didn't know the boy and didn't want to send someone to Stanford who was not a strong student. I gave their son a proficiency test, of sorts, to see what he was capable of, and he did not impress me. So I made an arrangement for him to study for half a year in Beijing, supervised by people I knew. If he did well, I promised to recommend him for admittance to Stanford.

I thought that was a pretty fair offer, but my hosts didn't agree. They found another way to get their son to the States, where he eventually, and coincidentally, became Rick Schoen's student. But this young man never became as good a mathematician as he might have been, in my opinion, because he wasn't that invested in learning about the subject. This seems to me to be a cultural problem: Many Chinese students don't make learning their top priority during graduate studies. For many of these young people, money is the primary objective whereas education and an overall appreciation of the subject are secondary, at best. They focus on some small thing mathematically and get a discrete result that they can publish in a paper, seeing that as a step toward advancing their careers and eventually commanding bigger salaries.

During this trip and others I later took to China, I met with what seemed to be a whole generation of young mathematicians, or would-be mathematicians, who lacked proper training and, in my opinion, lacked proper motivation as well. Many people got angry with me if I did not immediately recommend these students for graduate programs in the United States, even though I knew they would not be able to pass the qualifying exams. Those encounters grew tiresome, but the requests kept rolling in.

From Canton, Wang and I drove a van to Meizhou, the city where my mother had been born. I met some relatives that night and left the next morning for Jiaoling. The drive on the unpaved road took about an hour and a half. It was covered with yellow sand that looked brand new, which struck me as rather curious. To the explorers in the *Wizard of Oz*, "Follow the yellow brick road" was their guiding motif. In this case, we were following the yellow sand road, which puzzled me, as I'd never seen a thoroughfare like that before.

A couple of years later I got to the bottom of that mystery: The road with the freshly poured sand was brand new; it had been constructed just for my visit. I felt guilty that someone had gone to all that trouble for me, who, in the general scheme of things, wasn't that big of a deal—a thirty-year-old guy of Chinese descent who was little known outside the rarefied world of differential geometry. But I finally understood why I had been getting so much resistance, as well as a good deal of stalling, with regard to my initial request to visit Jiaoling. Someone, somewhere, was trying to hold me off until a thoroughfare could be built.

The town was quite primitive. It did not have a hotel, which was not surprising, given that until a short while before, no serviceable road reached it. I was put up at a guesthouse that had far more mosquitoes than guests. Mosquito netting covered the bed, but that merely served to keep the biting creatures close by. I was buzzed and attacked all night long. A loud bell, located right outside the house, rang loudly at 5 a.m., waking up everyone in the vicinity except for me, as I was already up and had been for most of the night.

The next day I went to the tomb where my grandfather and other ancestors were buried. (My father was buried in Hong Kong.) I next saw the home where my father had been born and in which he and my mother once lived. The place was run down and had a dirt floor, or more accurately, a mud floor.

Many relatives joined me during this tour; they were people I didn't know, but I got the distinct impression that they were expecting me to invite them to lunch, which I did. They slaughtered a cow in my honor, which cost me the not quite princely sum of 300 yuan (only about $14 back then). In China, in those days, you weren't supposed to kill a cow outright. You first had to say it was disabled; then it was okay to kill it. When it came time for eating, I was given a piece of meat that was a solid slab of unadulterated fat. That was considered the prime cut, and since I was regarded as the most important person there (as well as the person footing the bill), that piece was reserved for me. It looked unappetizing, to say the least. I poked at it a bit, unsure of what to do with it.

A lot of kids were running around barefoot. They seemed to be having a good time, as kids often do when allowed to run free, but they were poorly clothed and didn't appear to be in the best of health. I had 200 yuan in my pocket after paying for the cow. I started giving my relatives 10 yuan each, but then, as more of them showed up, I reduced that to 5 yuan, and then to 1 yuan, until I had nothing left. The fact that some people got more than others, and some got nothing at all, created a fight among the villagers. People made many other requests for help from me, most of which I was unable to oblige. And that led to more resentment.

The net result was that my grand homecoming did not quite live up to my lofty expectations. Some of that was because I had a romanticized picture of country life in China that did not accord well with the stark realities of the poor, struggling nation I was seeing. But I was also dismayed by some aspects of a Chinese culture that places too much emphasis, and too many expectations, on relatives. There are certainly some advantages to this tradition, because relatives can help one another during times of hardship. But the pendulum can, and often does, swing too far in that direction.

In the United States, many people realize that some things simply can't be done—that some things are too much to ask for and therefore should not be pursued. But in China, it seems to me, there are far fewer restraints. Many people feel that if you are a relative, you are obliged to do whatever they ask of you, regardless of what is feasible or ethical. I've seen this played out on countless occasions—in family settings, as during my visit to Jiaoling, and in the halls of academia.

Coming up against that kind of mentality time and again has surely caused problems for me, but that mind-set has also caused problems for Chinese society and many of its institutions by creating a culture of dependence. Too many people lack initiative, waiting for others to do things that they ought to be doing themselves.

I returned to America with mixed feelings. I was gratified to have finally seen China and trod upon native soil, but the experience had been somewhat disillusioning. I saw that my homeland had a long way to go toward attaining the living and educational standards that are common in the West. A few years after the end of the Cultural Revolution—an era of widespread purges, mass killings, and almost unimaginable upheaval—China's economy was in poor shape. The Great Famine that had caused the deaths of tens of millions of people had occurred less than two decades earlier, and people in the United States were still talking about "starving children in China." That catchphrase was often directed toward American children as an admonishment not to waste food, and to eat their vegetables, but it also spoke to the broad array of problems that China faced.

The challenges were daunting, overwhelming, in fact, and I didn't have a clue as to what I—one person wandering about a land populated (back then) with nearly a billion people—could do about it. But I still had hopes of finding ways to make a contribution, using whatever influence I had to help out, at least a little bit. A finger in the dike, perhaps. But with enough people and enough fingers, the waters might be held back long enough so that some important things could get done and some crucial gains made.

In Praise of an Older Sister
We cherished the bonds of youth.
Heedless frolics through fields of green.
Carrying books and pretend swords, we scampered up hills,
holding hands and issuing loud, delighted squeals.

Who knew that in just a year or two,
those blissful times would so brusquely change?
Our father died. Our brother struck by disease,
leaving us to endure many long, difficult days.

While our loving mother carried the burden,
Shing-Yue's compassion knew no bounds.
Always placing our futures before hers,
until illness came to take her down.

None of us could hold back the tears, nor did we even try.
For we are made of flesh and blood, not metal and stone.
Hardly a day passes by when I don't look back in regret,
lamenting the sad event that ended our friendship of old.

—Shing-Tung Yau, 2007

A Visit to My Teacher
In early spring, upon a verdant Berkeley hillside,
I passed through the courtyard I'd visited many times before.
The views from up high, still peerless and unchanged,
familiar vistas unleashing a surge of fond memories.
Of rousing dinners and times of song, merriment, and games,
all courteously presided over by a gracious host and hostess.

Gazing at the bay before me and undulating ridges to the sides,
I recalled my early ambitions, some realized, some not.
I've always valued your kindness and your guidance along the way.
Age brings perspective, though my aspirations have never waned.

—Shing-Tung Yau, 2001

Wedding photo of my parents,
Yeuk Lam Leung and Chen Ying Chiu, 1941.

The Yau/Chiu family in Shatin, Hong Kong, 1955. *Front row from left:* S. T. Yau, Shing-Yuk, Yeuk Lam Leung (mother) holding Shing-Ho, Chen Ying Chiu (father) holding Shing-Toung (Stephen), Shing-Kay, and Shing-Yue. *Back row from left:* Shing-Shan and Shing-Hu.

I taught tai chi to the mostly foreign professors at Chung Chi College (1968).

My graduation from Chung Chi College, 1969.

My first airplane ride, en route to the United States in 1969.

At the University of California, Berkeley, 1969, shortly after my arrival.

In 1979, upon landing at the Beijing Airport, I was in China for the first time since my infancy, some thirty years earlier.

Visiting the Summer Palace in Beijing, 1979.

I gathered with relatives and townspeople in Jiaoling, China, in 1979.

Standing in front of my ancestral home in Jiaoling, China, with two locals (1979).

In 1980, I toured the Great Wall of China with Raoul Bott (*right*) and Lars Gårding (*background*).

My wife, Yu-Yun, and me at the Forbidden City, Beijing, in 1980.

My friend and collaborator Bill Meeks at the University of California, Berkeley, 1981. (Photographer: George M. Bergman. Source: Archives of the Mathematisches Forschungsinstitut Oberwolfach.)

I received the Fields Medal in Warsaw, 1983.

In 1984 I met with Yaobang Hu, the general secretary of the Communist Party (*right, seated*), in the Great Hall of the People in Beijing.

At home in San Diego with friends and students, 1987. *Front row from left:* Weiyue Ding, Sheng Gong, Ta-Tsien Li, me (with my son Michael), and Xiao-Song Lin. *Back row from left:* Conan Leung, Gang Tian, and Shiu-Yuen Cheng.

A busy period at Harvard in 1988.

Yu-Yun, me, and our sons Isaac (*left*) and Michael (*right*) while sightseeing in Seoul, South Korea, 1992.

Shiing-Shen Chern, my former advisor, and me in 1996 at the Academia Sinica in Taiwan.

Flanked by two Nobel laureates in physics, Chen Ning Yang (*left*) and Samuel Ting (*right*), at Tsinghua University in Beijing, 1996.

In 1997 I received a National Medal of Science in Washington, D.C., from President Bill Clinton. James Watson is second from the left, and Robert Weinberg is third from the left. (Photo courtesy of Carol Clayton.)

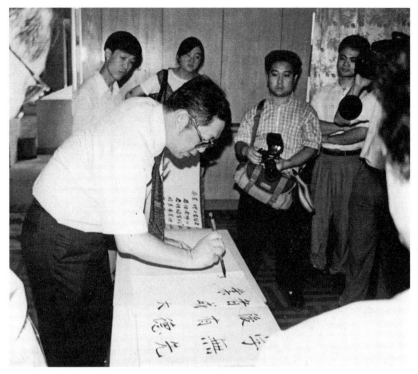

Demonstrating my calligraphy techniques before students in China's Xinjiang Province, 1998.

Richard Hamilton and me in Hangzhou, China, 2001.

Eugenio Calabi with me at Harvard in 2002.

Stephen Hawking, me, and Hawking's former student Zhongchao Wu on West Lake in Hangzhou, China, 2002. (Photo courtesy of the Center of Mathematical Sciences, Zhejiang University.)

In 2006 I met with China's future president, Xi Jinping, in Hangzhou, China.

With my longtime friend and collaborator Richard Schoen in 2012.

CHAPTER SEVEN

A Special Year

SYNERGISM OCCURS when the interaction of two or more agents yields a combined output that's greater than the sum of the inputs, or it may lead to an outcome that the individual agents simply could not have achieved on their own. Synergistic phenomena are ubiquitous in the natural world. Two hydrogen atoms, for instance, can combine with one oxygen atom to create H_2O, or water, which covers 71 percent of the surface of our planet, possessing seemingly magical properties—including the ability to sustain life itself—that the separate ingredients lack. Working together, colonies of bees and ants can accomplish tasks that single members of the species could not even attempt. An individual neuron cannot do much at all, but one hundred billion neurons, linked through one hundred trillion synaptic connections, collectively form the human brain, which is capable of feats that technological devices made by humans cannot mimic nor even approach.

Synergistic effects often arise in human interactions too—such as the bucket brigades of the mid-1600s that formed to douse fires in New Amsterdam, the colonial city subsequently renamed New York. Some 350 years later, I was hoping that the combined brainpower we were assembling at Princeton for our program in geometric analysis could tackle problems of a more intellectually demanding nature, though perhaps of lesser life-and-death urgency.

History shows that the greatest breakthroughs in mathematics have been made by solo practitioners and small groups of people; important problems are rarely, if ever, solved by committees in which duties are parceled out piecemeal, like homework assignments. Nevertheless, I still believe in the value of bringing together smart people, who work in different though overlapping branches of mathematics, to facilitate the exchange of ideas, while also giving those same folks the space and resources to pursue their interests without competing demands on their time. My work has always benefited from being in such an environment, and I felt there was a good chance that the eight months from September 1979 through April 1980 could prove to be exciting and eventful. I was doing everything I could to ensure that the "special" year on geometric analysis at IAS would live up to the boastful adjective associated with it.

I invited a number of outstanding researchers, and almost all of them came for some or part of the program. Among the core people who attended were Eugenio Calabi, S. Y. Cheng, Rick Schoen, Leon Simon, and Karen Uhlenbeck, along with Jean-Pierre Aubin, Jean-Pierre Bourguignon, Robert Bryant, Doris Fischer-Colbrie, and Peter Li. Several of my graduate students came, including Andrejs Treibergs. Enrico Bombieri, a Fields Medal winner on the IAS faculty, participated as well. And we had shorter-term visitors, too, such as Jeff Cheeger, Stefan Hildebrandt, Blaine Lawson, Louis Nirenberg, Roger Penrose, Malcolm Perry, and Yum-Tong Siu.

This was, according to Armand Borel, the biggest special program in mathematics that IAS had ever hosted. Although he was the IAS faculty member overseeing the workshop, he mostly let me do as I pleased. I decided to hold three seminars a week—one on differential geometry, one on minimal surfaces, and one on general topics, with an emphasis on general relativity (led by people like Penrose and Perry, a former student of Stephen Hawking) and other areas of mathematical physics. Borel said that the level of "cooperation between mathematicians and physicists [was] probably a first here since the early days" of the Institute.

I invited almost all of the speakers, many of whom were already attending the special year. Just as I'd hoped, we established an atmosphere during the program in which ideas flowed freely and unimpeded. People were motivated to work hard because they were passionate about the subject, not because they were under any pressure to do so. I was pleased at all the research that got done that year, much of which was presented at the program seminars. I provided an overview of geometric analysis to start things off. Calabi talked about some of his recent work on Kähler manifolds—the kind of spaces that lie at the heart of the conjecture named after him. Bourguignon and Lawson explored some geometric aspects of Yang-Mills theory. And Penrose discussed some unsolved problems in classical general relativity that were of particular interest and relevance to geometers. Schoen and I, meanwhile, proved a variant of the original Poincaré conjecture—this one involving noncompact surfaces (or manifolds) in which the Ricci curvature was positive.

Math was, of course, the main priority of everyone there, but we also set aside time for fun, creating what would be called today a work-life balance that contributed to good spirits and, I would wager, overall productivity. We went out to eat frequently and met every Saturday morning to play volleyball. We also played ping-pong. Bombieri was a much better player than I was, but he was not good enough to beat Simon. Each time Bombieri lost, he came up with a new excuse, attributing the setback to a sore arm, a stiff wrist, or some other ailment.

Qi-Keng Lu—the deputy director (under Loo-Keng Hua) of the Institute of Mathematics at the Chinese Academy of Sciences—came to IAS for several weeks during the special year. Lu had made some noteworthy contributions in the field of several complex variables. As one of Hua's leading students, Lu had also done his share, unfortunately, to perpetuate the fight between Hua and Chern. But Lu had played a big role in organizing my "homecoming" trip to China in 1979, and I wanted to return the favor by showing him New York City. Cheng and Siu, who were more up on things to do "on the town," took the lead during that tour.

We walked around 42nd Street, taking Lu to see the stage performance of *Oh! Calcutta!*, which featured many scenes involving nude women and men and the things they might do together when unfettered by clothing. Public displays of that sort were unheard of in mainland China (and caused a certain level of controversy at the time also in the States), and I was worried that Lu would be offended by the show. But I was relieved, and surprised, to discover that he was very much entertained by that long-running Broadway hit.

Back in the math quadrant of the normally staid IAS, some big parties were thrown, with lots of drinking and dancing, and I was told that the best of these affairs, coincidentally or not, were held in my two-bedroom apartment when I was out of town. After all, people tend to cut loose when the so-called boss is away.

One of the trips I took in the fall of 1979 was to Cornell, where I'd been asked to give a talk. I was also eager to visit Richard Hamilton, a Cornell mathematician who did not participate in the IAS special year but had embarked on a very intriguing, and exceedingly ambitious, project related to the notion of "Ricci flow." Flow, in geometry, involves changing the shape of a space or surface through small, continuous steps. One could, for instance, use a pump to slowly transform a deflated basketball into an almost perfect sphere. Or one could do something similar via mathematics, driving the shape-changing process through differential equations instead—equations that are, at their essence, about incremental (as in infinitesimally small) changes. The technique pioneered by Hamilton, Ricci flow, offers a way of smoothing out large-scale or "global" irregularities in complicated spaces and surfaces so that their overall geometry becomes more uniform. The process can, however, create small-scale or "local" irregularities, and the key challenge of the Ricci flow approach is to understand those irregularities when they crop up and figure out how to handle them or prevent them from forming in the first place.

It was a fascinating idea, but the relevant differential equations, which came to be known as the Hamilton equations, were very difficult to work with. I wasn't sure at first how to overcome those difficulties in order to make this method truly useful, but Hamilton was undaunted, sticking to his agenda for the next several decades, and he soon made impressive strides. I kept close tabs on this work over the years, touching base with him when I could and regularly arranging for my graduate students and postdocs to work with him.

Several things happened in the spring of 1979, apart from the program under way at IAS, which helped make it a "special" year for me. I was named the California Scientist of the Year—the first mathematician, and the youngest person, to receive this honor in the award's twenty-plus years of existence. My friend Michael Steele, with whom I'd spent a lot of time when he was a Stanford graduate student, urged me to buy a tuxedo for the awards ceremony rather than renting one. This, he said, would be the first in a series of important prizes given to me. I took his advice and bought a tuxedo, which I used a couple of times. But I soon gained weight and could not fit into it anymore.

At first, I was not particularly enthusiastic about the Scientist of the Year award, as I'd never heard of it before. I even told Steele that the decision of a small selection committee has no real bearing on the value of a person's work. History, I said (perhaps somewhat pompously), is the only true judge. On the other hand, my mother, who attended the awards ceremony with her California-based cousins, was extremely happy about the prize. And that made me happy too, in light of the fact that she had worked so hard, for so many years, to raise me—and bring me to a position where I might achieve some prominence.

Another thing of note that happened late in 1979 was that Borel popped unexpectedly into my office to tell me that Harvard would soon offer me a job (which turned out to be true), but I should wait before accepting because IAS would offer me a job too (which turned out to be true as well). I also heard from friends in Hong Kong that I would be receiving an honorary degree from CUHK in the following year, which was welcome news, considering I never got a bachelor's degree from that university, my alma mater.

However, other news coming from Hong Kong was not good at all. My older brother Shing-Yuk, who'd been battling brain cancer for about a decade, had taken a turn for the worse. He'd been working in a grocery store but had to be hospitalized when his condition began to deteriorate. X-ray scans revealed a tumor deep within a central part of his brain, and surgeons didn't know how to treat it. I spent two weeks in Hong Kong in December visiting him, which gave me a chance to see the kind of care he was getting (and not getting).

Neither of us was satisfied with the doctor assigned to his case. After his physician refused to share the medical records with a surgeon whom we liked, I resolved to take my brother to the United States for treatment, which was easier said than done. Shing-Yuk's first visa application was turned down. I sought help from IAS officials, who asked a New Jersey congressman to intervene on our behalf, but that didn't go anywhere either.

I then appealed to Andrew Tod Roy, the vice president of Chung Chi College, whose son, J. Stapleton Roy, was a senior diplomat who later became the U.S. ambassador to China. Andrew Roy wrote an impassioned letter on our behalf, but the embassy decided against issuing a visa to my brother.

Fortunately, my friend Isadore Singer was a presidential science advisor at the time. He played tennis with a very high-level official in the U.S. State Department, and with the help of Singer's friend, I was able to secure a visa for my brother.

Around this time, IAS offered me a permanent position, just as Borel had predicted. I faced a difficult decision because I loved Stanford and was also extremely impressed by Harvard. When I met with Raoul Bott, Heisuke Hironaka, David Mumford, and others there, I felt that I'd rarely been in the presence of so many intelligent people at the same time. IAS, of course, had a storied history too, as well as its own high-powered faculty. It had long been considered among the best places, if not the best place, for mathematicians to call home, owing to the priority given to research at the Institute and the volume of impressive work that had been carried out there.

One of the reasons I decided to stay at IAS was that many eminent members of the mathematics faculty—including Borel, Harish-Chandra, John Milnor, and Atle Selberg—made me feel quite at home there. Additional motivation stemmed from the fact that the Institute's director, Harry Woolf, had previously been provost of the Johns Hopkins Medical School, and I was assured that he could help my brother go to the Johns Hopkins Hospital in Baltimore. What's more, Johns Hopkins's celebrated director of neurosurgery, Dr. Donlin Long, was willing to treat my brother, in part because my brother's case was interesting to him and fit into his research program. Shing-Yuk's treatment, moreover, would come at virtually no cost. This was too great an opportunity to pass up, and I'm still thankful that Dr. Long was willing to waive his normal fees. I arranged for my brother to come to the United States as soon as his visa permitted him to travel, which turned out to be late in the summer of 1980.

Before that happened, the special year on geometric analysis was wrapping up in April, and several participants in the program encouraged me to present a list of open problems in the field. A decade earlier, after I had finished my first year of graduate school, Chern had gone to the 1970 International Congress of Mathematicians in Nice, France, where he discussed a number of unsolved problems with the potential to crack open new areas of mathematics. I vividly recall Chern telling me at the time that doing this sort of thing was one of the best ways of making a contribution to other researchers in the field. I also remember the quotation from the American inventor Charles Kettering, who said, "A problem well stated is a problem half solved."

With those words in mind, I ended up posing 120 problems, many of which I discussed at length during a series of lectures at IAS. I came up with most of these open problems myself, although some were contributed by other people or taken from the literature. But all of the problems were soon widely circulated, becoming known to practically everyone who did anything related to geometric analysis. About thirty of the problems have been at least partially solved, and the others have given people plenty to think about. I am not deluded enough to imagine that these problems, confined to a rather narrow sector of geometry, have been anywhere near as influential as the twenty-three mathematical problems that David Hilbert famously posed in 1900. But my problems did pique interest and activity in geometric analysis, and for that reason I believe that unveiling them at the conclusion of the IAS program was a fitting way to cap off the year.

That summer, after the IAS program had finished, I spent a couple of carefree months with my wife in San Diego. The two of us then went to China in August 1980, as I had agreed to attend a conference Chern was running in Beijing. Yu-Yun and I would also see some relatives and do some touring. Afterwards, I would continue on to Hong Kong to take my ailing brother to the United States.

The symposium Chern had organized on differential equations and differential geometry was held at the Friendship Hotel in Beijing. A number of important people—such as Bott, Lars Gårding, and Lars Hörmander—were there, as well as Chern, of course. It was an expensive affair for China to host, given how poor the country was at the time, but Chern hoped it would introduce Chinese students and researchers to some exciting ideas in geometry. He also recognized China's urgent need to send its students and scholars abroad, and for that reason connections with potential foreign hosts—such as Murray Protter, the head of Berkeley's Center for Pure and Applied Mathematics, who was invited to the symposium—was crucial.

In keeping with the lofty status of this event, the van that took my wife and me and other attendees from the airport to the hotel drove down the middle of the road to show that we were more important than the average vehicle and that others better get out of our way. Fortunately, not many cars were on the road back in those days, though a large number of bicyclists did have to clear a path, encouraged by the continuous beeping of the car's horn.

I lectured on the open problems I had raised a few months earlier at IAS, hoping to spark some interest among Chinese mathematicians, which eventually did happen. Chern had arranged some nice sightseeing trips in and around Beijing. My visit was marred, however, by another unpleasant encounter with Wenjun Wu's protégé, who again urged me, in an extremely combative manner, to endorse him for a major government award. When I refused to do so, our disagreement escalated into a heated altercation that aggravated my blood pressure condition, almost causing me to faint. After that stressful episode, the elder Chinese mathematicians who served as local hosts for the event tried to make sure I was not again disturbed by unexpected and uninvited guests.

But I was disturbed by something Chern said to the group of ten math "dignitaries" staying at the hotel. He called a meeting, ostensibly for the purpose of soliciting impressions from these experts on the state of mathematics in China, but he had a hidden agenda. He criticized the Institute of Mathematics, which Hua was heading, and urged that it be shut down, even though that was where the major work in the field was being carried out in China. Chern then asked the ten of us to write a letter to the Chinese government recommending that the institute be permanently closed. After his plea was met with dead silence, Chern repeated it.

I finally spoke up, saying that we were guests of the country, and it was neither our place, nor our business, to make such a request. Bott agreed with me, and the others quickly followed suit. They wanted nothing to do with this proposal. But Chern was furious with me, and my forthrightness contributed to further erosion of our relationship. But I don't regret taking a stand. A letter along the lines that Chern suggested, written by prominent outside scholars, would have been very damaging to Hua and the Chinese Academy of Sciences, as well as a disaster for Chinese mathematics as a whole, which already lagged far behind the West.

I assume that Chern's motivation stemmed from his long-standing feud with Hua, which, like many of these disputes, seems to have started for no good reason but was sustained by inertia. Although I was a student of Chern—and admired him greatly and was indebted to him in countless ways—I had nothing against Hua. I had learned many things from Hua as well, having reveled in his books as a child, and I was not aware of anything inappropriate that he'd ever done. I wasn't about to sacrifice Hua, and possibly cause permanent harm to the Chinese math establishment, just to please my former mentor. Nor did I feel the need to choose between Chern and Hua. Both were great mathematicians who had earned their places in the pantheon, and I never saw it as an either-or proposition. Mathematics is not a zero sum game.

Looking back on this incident, I doubt that Chern's ploy—ostensibly to harm the Chinese Academy of Sciences and Hua, the founding director of its Institute of Mathematics—came out of a vacuum. I believe he was responding to a 1977 report, published by the U.S. National Academy of Sciences (NAS), on the status of Chinese mathematics. It assumed the form of a short book, coedited by University of Chicago mathematician Saunders Mac Lane, who led a delegation of other American mathematicians to China a year earlier. The Pure and Applied Mathematics Delegation, headed by Mac Lane, made special mention of Jingrun Chen's work on the Goldbach conjecture and the Waring problem, as well as the work on value distribution theory by Yang Lo and Zhang Guanghou—all of whom were based at the Institute of Mathematics. That document had a deep impact on China, where various books, including primary school tracts, instructed people that they needed to learn from wise scholars like Uncle Chen, Uncle Yang, and Uncle Zhang.

Chern, I surmised, was not happy with this rendering of history and hoped to counteract the NAS report with the letter he'd drafted, cosigned by the ten noted mathematicians he had invited to Beijing, which would arrive at an opposite conclusion regarding the merits of the institute and its personnel. But much to Chern's dismay, this group, of which I was a vocal member, refused to play ball.

My time in Beijing was not taken up solely by mathematics and political maneuverings. Yu-Yun and I met with some relatives who were eager to move to the United States. We weren't able to help them, though we fielded many requests of this sort during our trip.

Soon after the conference ended, Yu-Yun and I traveled to Shanghai, where we joined thousands of couples who wandered aimlessly along the banks of the Huangpu River, a branch of the Yangtze that runs through the center of the city. It was a curious spectacle. Most of these people had nowhere else to go because they couldn't afford to eat in a restaurant. And even if they did have the money, one needed a permit or special coupon in those days—immediately after the Cultural Revolution—in order to purchase food at many restaurants. But Yu-Yun and I liked walking, so we were out with the rest of them on the famed waterside known as the Bund, strolling the banks of the scenic Huangpu, watching all the others doing the same.

Our next stop was Hangzhou, about one hundred miles southwest of Shanghai, where we took a boat cruise on the picturesque West Lake and saw some famous temples that had been badly damaged during the Cultural Revolution. Widespread destruction like this was a hallmark of that violent and tumultuous era. A couple of decades later, many of those beautiful, historic buildings were removed and replaced by unsightly, concrete structures.

Yu-Yun was pregnant during our trip and just starting to "show." As she was also starting to experience morning sickness, she decided to head straight back to San Diego while I made my way to Hong Kong. An official at the American Consulate, where I went to get the visa for Shing-Yuk, told me he was reluctant to let my brother leave Hong Kong. There was, in fact, an inch-thick stack of documents that made the case for why he should *not* be allowed to go. These arguments were countermanded, however, by an order that came from "high up in the State Department." I assumed that I had Singer and his well-placed friend to thank for that. And if that was indeed the case, I am grateful that Singer had taken up tennis rather than cricket or croquet.

The plane tickets were expensive because I had to get three seats in a row so Shing-Yuk could lie down. We flew to San Francisco first and then to Chicago. Our mother joined us on a plane from Chicago to Baltimore. Bun Wong, a mathematician I knew from high school who was visiting Johns Hopkins, met us at the airport and drove my brother to the hospital. My mother and I found an apartment somewhat near the hospital, but it was not in a great neighborhood. Even though she didn't speak any English, my mother somehow figured out how to take the city buses to the hospital each day so that she could be with her son. I had to get back to Princeton almost immediately, as my term at IAS was about to begin.

I returned to Baltimore soon for my brother's surgery, performed by Dr. Long, which lasted about ten hours. It was a complicated procedure because the tumor was located in the center of his brain. The recovery took a long time, but my brother was eventually able to walk a little, although his balance always remained a bit off. He had to wear a helmet at all times to protect his head, given that part of his skull had been removed.

When Shing-Yuk was finally discharged from the hospital, I brought my mother and him to stay in the home I had purchased on Locust Lane in Princeton—an ordinary suburban street back then, which is now part of a much pricier neighborhood.

My mother spent an inordinate amount of time at home because my brother needed almost round-the-clock care. To help her pass the time, some friends of mine, such as S. Y. Cheng and Bun Wong, occasionally came by to play Mahjong with her. I joined in when I was around. This sounds innocent enough, but years later a series of attacks were launched against me on the World Wide Web in which I was accused of forcing my students to play Mahjong with my mother. Those allegations—which seemed to be correlated with my questioning of the ethical behavior of a former student—were simply untrue. These people came of their own volition, just to be kind, and they were adults, not students. It seems bizarre that I'd even have to defend something like this, involving a board game with 144 tiles that's played by an estimated one hundred million people worldwide.

Back in the real world, where I prefer to reside, Borel was putting some pressure on me to edit the papers from the special year seminars and sort them into two books for Princeton University Press—one devoted to differential geometry and the other to minimal surfaces. As it turned out, I had mostly edited both volumes, and they were almost ready to go. In fact, I had written most of the sixty-page survey paper on geometric analysis in the waiting room of the Johns Hopkins Hospital. The first volume on differential geometry, which I edited, was printed in 1982. I let Bombieri take over the editing of the second volume on minimal surfaces, which came out two years later.

Around that time, I embarked on another big foray into the world of mathematics editing and publishing. In 1980, I agreed to become editor-in-chief of the *Journal of Differential Geometry* (*JDG*), taking over from the founding editor, Chuan-Chih Hsiung—a Chinese-born mathematician and friend of Chern's who was then based at Lehigh University.

JDG, which came into existence in 1967, was the first journal devoted to a subdiscipline within mathematics, as opposed to covering the field as a whole. The journal got off to an auspicious start, featuring papers by Marston Morse, Michael Atiyah, Isadore Singer, John Milnor, and other heavy hitters in the field. In fact, the Milnor paper, "A Note on Curvature and the Fundamental Group," which had made such a huge impression on me during my first year at Berkeley, was published in 1968 in *JDG*'s second volume.

The journal wasn't doing so well, however, when I was approached. Although I was up on the subject of differential geometry, and had been offered the editor's post for that reason, I had no experience running a mathematics journal. I was therefore hesitant to take on the job and agreed to do so only after Chern, Calabi, and Nirenberg all encouraged me. Hsiung wisely felt that I could use some help in this endeavor, suggesting that Phillip Griffiths and Blaine Lawson also come on board as editors.

I'd always made a point of trying to keep up with major developments in mathematics, and within the realm of differential geometry in particular, and I now had additional motivation for doing so. I was continually on the lookout for papers that might be a good fit for *JDG*. Because I spent time with Yu-Yun in San Diego each summer, UCSD had given me an office I could use about a quarter of the year, and I got to know Michael Freedman during these visits. Freedman, who was then a young faculty member, was trying to prove the four-dimensional Poincaré conjecture, which we spent a lot of time talking about, sometimes in or alongside the swimming pool in his backyard.

A group of topologists at Princeton University didn't think much of the method Freedman was pursuing, instead favoring the surgical techniques that John Milnor had introduced. I, however, was intrigued by Freedman's approach, which involved something called "Bing topology." When his work was sufficiently mature, I asked him whether we could publish his paper in *JDG*. Freedman agreed.

The Princeton folks soon realized they were missing the boat. They argued that the paper should appear in the *Annals of Mathematics,* published out of Princeton, which they considered the best publication under the sun. The Princeton topologist Bill Browder and his colleague Wu-Chung Hsiang called me, saying that it only made sense: The best papers in topology should appear in the best journal, namely the *Annals.* I was unconvinced, calmly explaining that Freedman and I had talked many times, after which he had decided on *JDG.* If he chose to withdraw his paper, however, I would let him do so, no questions asked. But I did make one last lobbying pitch, telling Freedman that his paper would be a big deal for *JDG;* it would give a big boost to the journal and, in so doing, could boost the field of differential geometry as well.

I believe that argument helped win him over, for in the end, Freedman stuck with *JDG,* and his paper—"The Topology of Four-Dimensional Manifolds," for which he later won a Fields Medal—was published in 1982. That did not endear me to the Princeton/*Annals* crowd, even though I, too, was then based in Princeton at IAS, only about a mile from the university.

Even Robion Kirby, a Berkeley topologist who had no direct involvement in this matter, expressed his displeasure with me over the Freedman paper. Many topologists did not like it when problems in topology were solved by nonstandard methods. Kirby was one of those people who were very protective of their turf, trying—or so it seemed to me—to keep their field free of interlopers. I don't have a high regard for that attitude, which strikes me as small-minded and contrary to the true spirit of mathematics. Nevertheless, I have rubbed up against that mind-set on numerous occasions, and sometimes those exchanges can be bruising. But I refuse to let myself be held back by tradition, especially when traditional methods can't get the job done.

JDG published another important paper in 1982. This one, by Clifford Taubes, was related to Yang-Mills theory. A year later, the journal secured a big paper by Simon Donaldson, which eventually earned him a Fields Medal. In that same year, 1983, *JDG* published "Supersymmetry and Morse Theory" by Edward Witten, which turned out to be hugely influential as well, even though some differential geometers initially raised a fuss about it too. The three referees I had lined up to review Witten's paper also raised a bit of a fuss, as they all voted to reject it. I, as chief editor, decided to override their objections, and I'm glad I did—primarily because of the impact the paper has had on math and physics, as well as for the impact it's had on *JDG* itself. The journal I had inherited, which had been close to folding a few years before, had undergone a remarkable turnaround: It was now a major player.

That said, I'd come to IAS to do research, and my new editing responsibilities didn't get in the way of that. I was starting to acquire a group of talented graduate students, the first of whom was Robert Bartnik, who originally came from Australia. Jürgen Jost, who'd been Stefan Hildebrandt's PhD student in Bonn, was my first postdoc, and he was quite skillful as well. For the past twenty-plus years, in fact, Jost has been the director of the Max Planck Institute for Mathematics in Leipzig, Germany.

I also began a student seminar at IAS that made some of my older colleagues uncomfortable. They believed that the Institute should hold only advanced seminars on new research, but I took a different view, arguing that an educational component would be worthwhile. Members of the old guard also complained about noise from these spirited offenders, though the high-decibel zones were mainly confined to an area near my office where the students were clustered. I was reminded of how my neighbors in Hong King used to complain when my father held poetry lessons for his children and other kids living nearby. While there are plenty of things to get riled about, I don't regard an enthusiasm for mathematics or poetry among the younger generation as being legitimate grounds for complaint.

One of my graduate students, an American of Chinese descent, was a talented young man. His father had shopped around for an advisor and picked me, based on a recommendation he had gotten from Atiyah. Wu-Chung Hsiang, who was then Princeton's math chair, was not happy with this arrangement. "You come from IAS and take the best graduate student away from us!" Hsiang complained. I calmly replied that I hadn't picked this student and forced him to be my advisee; he had picked me.

That outburst was rather ironic, given that it occurred at a dinner held in the home of Joe Kohn, a Princeton mathematician who strongly supported the idea of my taking graduate students from the university. I discussed the incident with Borel afterwards, who told me he'd had similar experiences, noting that there's always some competition between IAS and Princeton.

But the student's admittance to Princeton hit a snag when he failed to pass the oral qualifying exam. I asked a professor on the exam committee what he did wrong and was told that he did not understand the link between symplectic geometry and mechanics. I mentioned this to the algebraic geometer Nick Katz, Princeton's graduate student advisor, who confessed that he wasn't familiar with that link either. As a point of historical fact, symplectic geometry (a branch of differential geometry) did have its origins in Newton's laws of motion, which formed the basis for classical mechanics. This connection arose from observations made in the 1830s by William Rowan Hamilton, who uncovered deep mathematical symmetries between an object's position and its momentum. Nearly a century and a half later, symplectic geometry had changed dramatically from its roots to the point where many people weren't aware of its ties to classical mechanics—in the same way that many people forget that Botox was first developed to treat an eye condition and Viagra was introduced to lower blood pressure.

I argued that this student should be given another chance. Joe Kohn and I gave him another oral exam, and this time he performed well. But he was so dismayed by his earlier failure that he went home for six months. Afterwards, he did return to Princeton to earn a PhD, and he has since gone on to a fine career.

Ngaiming Mok, who was born in Hong Kong, came to Princeton in 1980, just after receiving a PhD at Stanford under the tutelage of Y. T. Siu. We talked soon after he arrived at Princeton and began to work on some problems together, following up on some things that Siu and I had joined forces on during the special year. Siu and I looked at noncompact Kähler manifolds, difficult-to-understand spaces that stretch out to infinity. But we found a means of closing up these extended spaces in such a way that we could analyze their structure at infinity. Borel, Mumford, Jean-Pierre Serre, Carl Ludvig Siegel, and others had attacked problems of this sort by algebraic means. I, however, initiated the program to attack such problems by analytic means through the use of differential equations and various geometric techniques. And Siu and I solved the first important case involving spaces whose curvature is strongly negative.

A year or so later, Jia-Qing Zhong, a student of Hua's in China, arrived at IAS as my postdoc. I suggested some problems in complex geometry that he and Mok could work on, under my guidance, and they did quite well, coming up with some interesting solutions. But Siu, as I mentioned, was quite competitive with me. And once he found out I was helping Mok and Zhong, and sometimes working with them, he became very protective, asking me to no longer collaborate with Mok. That put a stop to my collaborations with Siu and his students that has persisted to this day. I wasn't happy with that outcome, as Siu is an excellent mathematician. We had done some good work together, and I would have been happy to do more.

Meanwhile, I heard from Siu about another matter, this time concerning Peter Sarnak, who had been a graduate student at Stanford of Paul Cohen, a Fields Medal winner. Sarnak had stayed on at Stanford, and Cohen was hoping to promote him to full professor very quickly, just a couple of years after he'd gotten his PhD, which was quite unusual. Siu, who was then at Stanford, wanted me to ask a Princeton number theorist what he thought of Sarnak's work. I didn't want to do this, as I didn't know Sarnak, nor was I especially knowledgeable about his area of number theory. But Siu called me several times, so I finally felt obliged to talk to the number theorist in question. As Sarnak was barely out of graduate school at that time, it was understandable that the number theorist did not seem to be overwhelmed by Sarnak's early contributions. I then passed on this offhand impression of Sarnak's work to Siu.

I subsequently learned that at a Stanford faculty meeting held shortly thereafter, it was somehow communicated that I was against Sarnak's appointment—although I never said anything of the sort. I had merely relayed another expert's preliminary impressions after being repeatedly pressed to do so. In the process, I had angered Cohen, with whom I'd been on good terms, and jeopardized my relationship with Sarnak, who had been told that his fate at Stanford rested in my hands. In time, Sarnak and I have come to treat each other professionally and respectfully. But I learned a valuable lesson through this incident—namely, that academic politics can be quite delicate and sometimes even treacherous. I've been more careful since then to avoid getting tangled up in situations that don't concern me, though I've been only partially successful at this.

On March 21, 1981, I caught a last-minute flight to San Diego as soon as I heard that Yu-Yun's and my first child was about to be born, somewhat earlier than expected. Fortunately, I made it to the hospital that night, about eight hours before our son Isaac emerged from the womb. Yu-Yun had gone through an extremely tough labor that lasted more than twenty-four hours. She was in great pain for much of that time but refused to take any drugs because she wanted our son to come out healthy, and he did. When the baby finally emerged, cried, opened his eyes, and looked around, both of us were happy beyond words.

I stayed in San Diego as long as I could before going back to IAS to finish up the last couple of weeks of the term. Fortunately, Yu-Yun's mother helped her take care of the baby—a rather large, chubby boy— until I was able to return to San Diego for the summer. Raising a child was new to both of us, though I was surprised at how patient I could be. In mathematics, I was always restless, eager to keep pushing forward. Yet I could spend hours doing nothing other than hold Isaac and be completely content (when he wasn't screaming at the top of his lungs). That feeling of serenity seemed mysterious to me, though perhaps that's because I went into mathematics rather than biology.

Of course, I still had to return to Princeton in the fall to fulfill my obligations at IAS. It turned out to be a busy and eventful year. The mathematician Karen Uhlenbeck came to IAS for three days, and we spent all of that time working nonstop on the mathematics associated with Hermitian-Yang-Mills equations—central components of the quantum field theories upon which particle physics now rests.

Hamilton also contacted me out of the blue to report on the first major breakthrough stemming from his research on Ricci flow: He had proved a special case of the Poincaré conjecture—one involving compact, three-dimensional manifolds with positive Ricci curvature. I was surprised because I wasn't sure that his approach would ever pan out. But this latest work was both beautiful and exciting, and his result was far stronger than the one Schoen and I had achieved two years earlier. It was as if he had found a key that unlocked a door that had never been opened before. I quickly realized that the direction Hamilton had been pursuing would be very fruitful.

I invited him to come to IAS to give a series of talks. We spent a lot of time discussing the potential of Ricci flow methods. I told him that these techniques could be used to prove the Poincaré conjecture in three dimensions—a high-profile problem that had been unresolved since the start of the twentieth century. The application of these same techniques, I said, could also solve Bill Thurston's geometrization conjecture about classifying three-dimensional topological spaces into eight distinct types. Thurston's conjecture was broad enough to include the three-dimensional statement of the Poincaré conjecture, so that proving Thurston would imply a proof of Poincaré as well. I immediately got three of my graduate students—Shigetoshi Bando (from Japan), Huai-Dong Cao (from China), and Ben Chow—to start working on questions related to Ricci flow.

Hamilton, who had come from Cornell, stayed for a week in an IAS apartment. At the end of his stay, the chief math secretary was livid because Hamilton had made a huge mess of the apartment, and it took a long time to clean up the place. On the other hand, he had given some wonderful talks, and collaborations between Hamilton, my students, and me picked up from that point onward. So, on balance, his visit would have to be called a great success. Hamilton may have posed some challenges to the cleaning and janitorial staff, but he had posed even more consequential challenges to the mathematics community, some of which were taken up by members of my group.

Jürgen Moser, who'd been so nice to me when I visited Courant in 1975, had since moved to the Swiss Federal Institute of Technology (ETH Zurich), inviting me to go there for two weeks in the fall of 1981 to give some lectures before the International Mathematical Union (IMU). My postdoc Jürgen Jost accompanied me on this trip. His presence was helpful, as he was from Germany, and I've never been able to get very far with the German language. In addition to the mathematics-related activities, I took some walks in the mountains with Jost, and the Swiss landscapes were sensational, as advertised.

One night, I was invited to dinner at a fancy restaurant in Zurich with Moser and the Indian-born mathematician Komaravolu Chandrasekharan, a founding faculty member of the School of Mathematics at the Zurich institute. Both Moser and Chandrasekharan were high-ranking people at the IMU; Chandrasekharan had served as president during the 1970s, and Moser would become president in the following year. Chandrasekharan urged me to sit at a particular seat in the restaurant, later telling me that several mathematicians who sat there had gone on to win the Fields Medal. I didn't know how to interpret that remark, though I suspected he was privy to information I was not aware of.

But I didn't dwell on this matter as I was soon caught up in a wide variety of things. The physicist Gary Horowitz became my postdoc in the fall of 1981, although in IAS parlance he was actually called my "assistant." Horowitz, who had studied with Robert Geroch at the University of Chicago, was interested in generalizing the positive mass conjecture, which Schoen and I had proved two years earlier. Soon after his arrival at IAS, Horowitz started working on this problem with Malcolm Perry, who was then at Princeton, although I didn't know about their collaboration initially.

While the concept of mass is straightforward in classical mechanics, it is much more complicated in general relativity—a consequence of the nonlinearity of the governing equations. For the most part, mass in general relativity can be defined only for isolated systems that are very far away, essentially at infinity. Moreover, there is not a single definition of the term "mass." Different definitions apply to different situations, and in some cases there is *no* agreed-upon definition. When you're talking about mass in Einstein's theory, you are necessarily wading into a murky realm.

The proof by Schoen and me applied to so-called "ADM" mass—named for the authors of this formulation, Richard Arnowitt, Stanley Deser, and Charles Misner—for which there is a rigorous definition that practically everyone accepts. Horowitz and Perry were trying to extend the positive mass theorem to include "Bondi mass," which is less clearly defined. Many physicists believe that the Bondi mass of a system is equal to its ADM mass minus the energy carried away by gravitational waves—the radiation associated with the force of gravity. Einstein predicted the existence of gravitational radiation in 1916, and this prediction was confirmed one hundred years later based on observations made at the Laser Interferometer Gravitational-Wave Observatory (LIGO).

The positive mass conjecture asserts that the energy of a physical system always remains positive. This means that the ADM mass, which Schoen and I had already shown to be positive, cannot be completely carried away by gravitational radiation. The Bondi mass, therefore, must be positive too, and that's exactly what Schoen and I were trying to establish.

As I've said, I had no idea that Horowitz was collaborating with Perry on this until my graduate student Robert Bartnik casually mentioned that they'd just about finished their proof. I was miffed to find out that my assistant had been doing this without telling me, but I used that news as inspiration for Schoen and me to finish the work we'd already done on the problem.

Schoen was then at the Courant Institute, and I joined him there early the next morning. We worked the entire day without stopping, finishing our calculations at 6:30 p.m. Then I suddenly remembered that I'd been invited to dinner that night as the "guest of honor" at the home of François Trèves, a distinguished French mathematician who was teaching at Rutgers. There was no way I could make it, as the dinner had already begun and I was more than an hour away from New Brunswick, New Jersey. My lapse was particularly embarrassing because Trèves had invited me about two months before and had reminded me of the event several times.

More than thirty-five years later, I still feel bad about that faux pas. But at that point, after offering my apologies by phone, there was nothing I could do but finish the work that Schoen and I had started. Our paper, "Proof That the Bondi Mass Is Positive," came out in *Physical Review Letters* a couple of months later, right next to the article by Horowitz and Perry, "Gravitational Energy Cannot Become Negative." These papers offered further evidence regarding the stability of our universe, as well as reassurance that it is not collapsing.

Does it seem odd that I was motivated in this case to compete with my assistant? I don't think so. In my experience, it is extremely common in the field of mathematics—and throughout all of science—to be spurred on when another person or group is making advances on a problem to which you've already devoted a good deal of effort. As long as you don't copy someone else's work, or do anything else that is unethical, competition is healthy in mathematics. Indeed, the field depends on it for a good deal of the progress that has been made.

At around the same time, I met Zhiyong Gao, a former student at Fudan University in China who had since come to Stony Brook, with the help of C. N. Yang, where he pursued a PhD that was supervised by Blaine Lawson. Gao and I worked together to solve an important problem related to manifolds with negative Ricci curvature that had puzzled geometers for a long time. The problem, expressed in elementary terms, concerned whether it was geometrically possible to construct a simply connected manifold (one with no holes coursing through it) that had negative Ricci curvature. Ricci curvature is associated with the "cosmological constant"—a factor added to Einstein's equations that is thought to account for the accelerated expansion of our universe since the Big Bang. Negative Ricci curvature, which would correspond to a negative cosmological constant, is consistent with an expanding universe, but one whose expansion is decelerating rather than accelerating.

Gao and I drew on some prior work by Thurston to produce an example of a manifold, a three-dimensional sphere, which possessed the desired geometry. I considered this a significant achievement and accordingly wrote Gao a strong letter of recommendation that helped him land a tenured position at Rice University.

To my dismay, it appeared that Gao's interest in research declined soon after he got tenure. His publications dropped off, from what I could tell, as did his attendance at math conferences. I've seen this happen with other Chinese students, who are eager to get a good job but don't seem to be that fired up about math itself. This may be an unintended consequence of the Chinese educational system, which tends to present subjects in a rote manner that can suck the life out of them.

While that was certainly disappointing, my next brush with China, or at least with a few if its people, had even more negative repercussions. It started innocently enough with a phone call from Chern, who was trying to help his friend, Shisun Ding, the chair of the math department at Peking University. Chern was hoping to expand Ding's influence in China. The president of Peking University would soon be stepping down, and Chern wanted Ding to take his place. But Ding first needed to beef up his curriculum vitae. He was hosted by Phillip Griffiths at Harvard at the time, which would make for a prestigious entry on his CV, even though Ding didn't seem to be doing much mathematics there. Chern asked me to get Ding an appointment at IAS.

I told Chern that it was too late, as I had already offered my assistant position to Gary Horowitz. That is the only appointment faculty members can make unless they can persuade their colleagues that someone is such a superb mathematician that IAS simply has to hire him or her. I could not, in good conscience, make such a case for Ding, who had not yet made any outstanding contributions to mathematics that I was aware of. Furthermore, Ding worked in algebra—a field that many people at IAS knew far better than I did—and I didn't feel comfortable trying to push through an appointment in that area. Nor did I have reason to believe that such an attempt would be successful.

Chern, of course, was not happy about my response. He believed that if I had wanted to do something for Ding, I would have, but I chose not to. That, in turn, made Ding angry at me. He became president of Peking University in 1984 and later served as chairman of the China Democratic League, one of the eight recognized political parties in the People's Republic. In this relatively brief encounter with Ding, which I had not wanted any part of, I had made a powerful enemy. Peking University, a leading Chinese university, soon became less friendly toward me—a fact that did not make my subsequent dealings in China any easier. This, in turn, was part of a broader power struggle between the mathematics group at the Chinese Academy of Sciences, led by Hua, and a faction at Peking University, which was under Chern's sway. I often found myself in the crosshairs of that bitter fight, which was not a pleasant, or particularly relaxing, place to be.

In April 1982, after finishing my term at IAS, I flew to San Diego to be with Yu-Yun and Isaac, with whom I'd celebrated his first birthday a couple of weeks before. While there, I got a call from my brother Stephen, who had received a letter from the IMU that had been sent to me at IAS, where he had a one-year appointment. This letter informed me that I had been named one of three winners of the 1982 Fields Medal for my work on the Calabi conjecture, on the positive mass conjecture, and on real and complex Monge-Ampère equations—the first person from China ever to receive this award. The other two 1982 winners were Alain Connes of IHES (for his work on operator algebras and other topics) and Bill Thurston of Princeton (for having "revolutionized the study of topology in two and three dimensions"). The prize ceremony was supposed to have been held in Warsaw, Poland, in 1982, as part of the International Congress of Mathematicians, but the IMU decided to postpone its meeting for a year, until August 1983, because martial law had been declared in Poland in late 1981 as part of the government's attempt to suppress the prodemocracy Solidarity movement. Fortunately, martial law was lifted in July 1983, so the IMU was able to go ahead with the event a month later.

Yu-Yun took a three-month leave of absence from her job so that she and Isaac could be with me during the fall of 1982, though she preferred to live in Philadelphia rather than Princeton. We found an apartment close to Eugenio Calabi's home, and he was kind enough to loan us a crib and other baby gear, even taking the time to help us set things up. It was an hour's drive to Princeton. I bought a run-down car for $200 that was in decent working condition but looked terrible. The secretaries at IAS thought it was disgraceful for a faculty member to drive a car like that and, worse yet, have the nerve to park it in the Institute lot.

Borel, who had always been extremely supportive of me, disapproved of my mixing business with pleasure by bringing Isaac—then a beefy toddler about a year and a half old—into the IAS dining hall, so I refrained from doing that again. Princeton was a proper place, and years of living on the casual West Coast had apparently left me unfit for the more rigorous social standards of the dignified East.

In April 1983, I went to Berkeley for three months to participate in a program on geometric analysis that Chern had organized. Schoen and I taught a several week–long course, discussing some new theorems we had proved, based on minimal surface arguments, regarding manifolds of positive scalar curvature. Several Chinese students at Stony Brook told me that a former student of Lawson's had taken detailed notes during the course. It looked as if those notes may have been passed on to Gromov and Lawson, because a preprint of their subsequent paper, which Schoen saw, seemed to incorporate some of our ideas. Schoen complained about this in a letter addressed to Lawson, which he deposited in a mail slot in Berkeley's Evans Hall. But the mail slot was blocked, and the letter was returned to Schoen a few months later. By that time, it was too late for him to send the letter, and the matter was dropped, as it probably should have been. Mathematics, after all, is a competitive business.

Our second son, Michael, was born in June 1983, which was, of course, another joyous event. The overpowering sense of bringing a new being into the world never gets old, and the emotional impact of the occasion hit me just as hard the second time around. Two months later, however, Yu-Yun and I had to leave Michael and Isaac in the care of my mother-in-law while we went to Warsaw for the Fields Medal ceremony. Demonstrations against the Communist regime were still going on, and Thurston advised me against saying anything to the reporters who were trying to interview us. That was fine with me, as I had a hard time understanding what the reporters—many of whom did not speak much English—were saying anyway.

After the award ceremony, Yu-Yun and I were invited to join Y. T. Siu, Wu-Chung Hsiang, and others for a drink. And that's where a topic of conversation that I had not anticipated set the stage for big problems down the road—problems for me in particular.

Siu and Hsiang were strongly opposed to a program Chern was formulating with Phillip Griffiths to bring Chinese students to North America. That plan was modeled after the China-U.S. Physics Examination and Application (CUSPEA), a famous program started a couple of years earlier by the physics Nobel laureate T. D. Lee to help Chinese physics students attend graduate school in the United States and Canada. In the wake of the Cultural Revolution, school transcripts, teacher recommendations, and the like were hard to come by in China, and Lee (who was then at Columbia University) worked with American physicists to design an exam that would help them pick out about one hundred Chinese students each year for placement abroad.

Chern was trying to do something similar for Chinese mathematics students, who greatly outnumbered physics students in those days largely because expensive experimental facilities were not needed in math. There was nothing wrong with the general idea Chern was espousing, and indeed much to recommend it, but Siu, Hsiang, and I were not thrilled about some of the particulars. Under the proposal Chern had worked out with Griffiths, students applying to the program in pure math would be given an exam by Griffiths; students interested in applied math would be given an exam by David Benney of MIT. Griffiths and Benney, acting on behalf of the American Mathematical Society (AMS), would then have a large say over who among these students would go to which school.

I was uncomfortable that this program with China, as it was then constructed, would place so much power into so few hands. Siu, Hsiang, and I all felt that the participating Chinese students should have more say over which schools they wanted to attend than this plan allowed. We preferred an approach that made it easy for these students to apply to American schools directly, thereby giving them more choice and reducing the control granted to the AMS.

I had asked Chern on three occasions whether AMS's role in the program was his idea, and each time he said no—it had nothing to do with him. In view of those statements, our questioning the AMS piece of the program should in no way have been construed as an attack on Chern, Griffiths, or Benney, as I had nothing against any of them.

Nevertheless, Hsiang, Siu, and I were concerned enough over this issue that someone finally suggested sending a letter to China's minister of education that would present our ideas on alternative ways of running the program. We never wrote that letter, but several months later I was with S. Y. Cheng, my graduate student Huai-Dong Cao, and Chang-Shou Lin, an IAS visiting mathematician originally from Taiwan, and we got onto the same subject again. This time, we drew up a rough draft of a letter along the lines I had previously discussed with Siu and Hsiang. As we would need the others' support, I initially sent a copy of the letter—an unfinished, handwritten draft—to Siu to get his reaction. From there, our draft somehow made its way into the hands of Griffiths and, shortly thereafter, into the hands of Chern.

Griffiths, unsurprisingly, was not enthusiastic about the letter—or I should say this extremely preliminary draft of the letter. Nor was Chern, who was up in arms over my apparent "treachery." He then complained to all his friends at Berkeley, Courant, Princeton, and elsewhere, tearfully proclaiming, "Yau has betrayed me!" Chern spread that message far and wide, and even people who had been supportive of me, like Moser and Nirenberg, were shaken up by his allegations.

This was another step in the protracted falling out between Chern and me—a period that lasted almost until his death more than twenty years later. I had stood up for what I believed was right, even though I knew my position might displease my mentor, which it ultimately did. But things turned out worse for me than they might have, in part because Hsiang and Siu—who also had big roles in initiating the letter—seemed content to let the blame fall squarely on my shoulders.

Looking back on this incident, it's kind of ironic that some of the grief that befell me, months and even years later, started with a casual conversation that took place during a time of celebration, just hours after I'd gotten the Fields Medal—widely regarded as the most prestigious award in mathematics.

Winning such a prize would normally be considered a joyous occasion, though in my case it was offset by another, rather grave development. For while I was in Warsaw, being feted and carousing with my wife and friends, my brother Shing-Yuk had gotten sick again. He was taken to the hospital where doctors found a blood clot in his leg, for which he was prescribed blood thinners. Shing-Yuk was kept on blood thinners for a while, too long as it turned out, which caused bleeding in the brain that eventually resulted in a coma. He stayed in a coma for the last six months of his life. It was a tragic end for my brother, who'd fallen ill at a young age and never had the chance to do all that he might have done. Of course, I had no idea things would unfold in this way while I was in Warsaw, but I still thought of him often and was preoccupied with concerns over his frail health.

I learned from this latest flap among colleagues, and many others over the years, that life is never just one thing. You can't keep going up—even after capturing a great honor like a Fields Medal. Eventually, gravity will overtake you and pull you back down, in some cases crashing down.

I have mixed feelings about mathematical awards, of which I've been lucky enough to win more than a few. I have never done any work in the field for the purpose of winning a prize, as I am of the opinion that doing mathematics is its own reward—especially when the work goes well. On the other hand, it is nice to be recognized for the hard work you've put in. But recognition—and fame, if you could call it that—does have its trappings. I was no longer an anonymous researcher who could move around unnoticed, focusing on mathematics twenty-four hours a day if I cared to. I was now somewhat of an authority figure, someone whose convictions carried weight. As such, I was asked to express my opinions and sometimes assume a bigger role in policy, administrative, and political matters, which inevitably resulted in my being drawn into skirmishes I had no desire to be involved in.

The news that I was the first person of Chinese descent to win the Fields Medal traveled fast, and I became something of a folk hero in China. But I also heard that some people were not rejoicing, or at least had mixed feelings about my award, if not actually feeling bitter. Perhaps they felt that they deserved a Fields Medal more than I did.

Someone else was angry at me too, though for an entirely different reason. My two-year-old son Isaac became upset every time I left San Diego to return to Princeton. He was starting to protest more violently, sometimes pounding the floor or even banging his head against it. While

I could handle the fact that some people in the mathematics world did not love me, and maybe bore some animosity toward me, I could not ignore the feelings of my own son, especially when they were displayed in such an emotionally raw and unfiltered fashion.

I was perhaps too slow in coming to this realization, yet it was now vividly clear to me that the present situation, living on the East Coast while my family was installed on the West, was untenable. I had to do something about it—to find a way of bringing my family together. Given that Yu-Yun did not want to come to Princeton, I needed to go somewhere else.

David Mumford had been coming to IAS regularly for a program in algebraic geometry he was running with Phillip Griffiths, and I told him I might have to leave. Mumford spread the word to Henry Rosovsky, Harvard's dean of the Faculty of Arts and Sciences, who came to Philadelphia (where I was then living) to try to persuade me to join the Harvard faculty. A charming and erudite man, Rosovsky even managed to weave a famous work of Chinese literature, *Romance of the Three Kingdoms*, into his argument as to why I should come to Harvard. Although I can't reconstruct the exact line of reasoning he put forth—and how he had related my employment at his university to the tale of three warlords struggling for dominance near the end of the Han Dynasty, some seventeen hundred years ago—I was still swept up by Rosovsky's rhetoric and smooth elocution. The bottom line, however, was that Harvard could offer only 75 percent of my current salary. That was a nonstarter, given our current economic situation: Yu-Yun and I now had two children to support, plus her parents and my mother and brother Shing-Yuk, who was still hanging on, though barely, at the time I was making this decision. So I reluctantly had to say no to Harvard for a second time.

I also felt bad leaving Stanford, where I'd been treated extremely well by Robert Osserman, the department chair Hans Samuelson, and many others. I had no complaints about Stanford at all, but the more I thought about it, the more it seemed that UCSD would now be the best choice, given that my wife and sons were already living in a home not far from campus. Isadore Singer, one of the most connected people I knew, put me in touch with his friend Richard Atkinson, the UCSD chancellor, who made a very tempting offer.

I then spoke with Borel, who kindly told me they'd keep my position at IAS open for two years in case I wanted to come back. But UCSD had some advantages that other schools could not match. First and foremost, my family was there, and Yu-Yun had a job in San Diego that she was happy with. Second, the university promised that I could make two additional appointments in the department so that I would have a handpicked team of colleagues to collaborate with—something I valued strongly. Rick Schoen agreed to move from Berkeley to San Diego, and Richard Hamilton agreed to move there from Cornell. With a strong group of us working on geometric analysis together, Hamilton believed that UCSD would provide "the ideal environment" for him to further develop his ideas on Ricci flow. San Diego offered Hamilton the ideal environment for another reason: He was an avid surfer and wind surfer and loved being near the ocean, and the UCSD mathematics building on Gilman Drive is not much more than a mile (as the crow flies) to the beach.

Although IAS had been a convivial home for me, one drawback was that it had been hard for me to get graduate students. I've always found that interactions with younger people are not only healthy, but essential. It keeps you current in the field. And with a constant influx of new students, your research is less likely to dry up. Lining up students at a large state university like UCSD would not be a problem.

Furthermore, with Hamilton, Schoen, and me, we had a good core of geometric analysts, and the German differential geometer, Gerhard Huisken, soon came to San Diego as a visiting professor too. I started thinking about bringing in other strong mathematicians and turning San Diego, which already had what was advertised as the world's greatest weather, into a mathematical paradise as well.

Michael Freedman was still at UCSD, after having gained considerable renown for his work on four-dimensional manifolds. Quite a few other accomplished faculty members were also there. With encouragement from the administration, I would soon get involved in building up the department further—without realizing at the time how much trouble that would prove to be, how much resistance I would face, or how many quarrels that would incite. Looking back, I realize I would have been much smarter to stick to my research. But sometimes you have to learn the hard way. And the hard way, for better or worse, has often been my way.

CHAPTER EIGHT

Strings and Waves in Sunny San Diego

BEFORE I MOVED from one end of the country to the other, as I was about to do in 1984, I needed to attend to a few things. One of them was going to China at the invitation of my friend Yang Lo, a prominent scholar who was soon to become the director of the Institute of Mathematics at the Chinese Academy of Sciences and later the founding director of the Academy of Mathematics and Systems Science. "Uncle Yang," as Chinese students affectionately called him, was a good person to know when traveling in China. He later accompanied me to the airport when I had just a few minutes to catch my flight, which I otherwise would have missed because the check-in desk was temporarily closed. He showed his ID card and the security guys saluted him. "You are Yang Lo," one said. "We read about you in our textbooks, and you and your friend can go right through!"

I took my mother along on this excursion, hoping that she could relax after the grueling death of my older brother. Lo arranged for us to meet a high-ranking official in the Communist Party. He was a nice man who regaled us with stories about fashion shows in Canton to show us how advanced China was becoming. Although I appreciated his casual style, I didn't think that Chinese culture would approve of a top official who spoke to the general layman in such an informal way. I suspected that he would not hold power for long, and sure enough, he was gone within a year.

My main goal for this trip was to do something nice for my mother, who was still in mourning after having taken care of my brother for so long. We took in some sights and saw some relatives, which seemed to boost her spirits somewhat.

But another objective for this trip was for me to meet with Chinese students in the hopes of finding good candidates to undertake graduate studies with me at UCSD. I wanted to try to give Chinese students the kind of break I received when I went to Berkeley at the age of twenty and the world of mathematics opened up for me. When I left for America then, in 1969, China was still in the throes of the decadelong Cultural Revolution—a period of widespread bloodshed and mass starvation during which university professors and other intellectuals were forced into manual labor and academic research ground to a halt. Conditions had improved by the mid-1980s, but China was still very poor, and its universities were far behind their Western counterparts. One of the ways I've tried to help the country is by bringing qualified Chinese graduate students, postdocs, and professors to top schools in the United States so they can be exposed to research at the highest levels and ideally participate themselves.

On this particular outing, I recruited Jun Li from Fudan University, Gang Tian from Peking University, and two students from the Chinese Academy of Sciences, Wan-Xiong Shi and Fangyang Zheng. All four of these students have experienced some success in mathematics, and all have served as professors at U.S. universities.

While in Beijing, I visited Loo-Keng Hua in the hospital where he was being treated for a serious heart condition. Hua was very frustrated because rumors were then circulating about an alleged affair that he vigorously denied. Hua assumed that his rivals were behind those rumors, but he did not explicitly blame Chern.

Hua sent a letter to me that alluded to this conflict, containing excerpts of a famous poem from the Tang Dynasty (circa 600 to 900 CE) about illustrious poets who were constantly competing over who was better at his chosen craft. "Only time will tell," the poem concluded, and it seemed that Hua believed the same statement might apply to the competition between Chern and him—a contest that neither seemed willing or able to relinquish to the end of their lives.

A little more than a year later, Hua traveled to Japan. While delivering a lecture in Tokyo on June 12, 1985, he collapsed on the stage, dying instantly of cardiac arrest. I was extremely distressed when I heard the news and thought of the longtime feud that likely contributed to Hua's heart condition. And even then, when Chern was the only one left standing, the fight persisted. The fact that I had met with Hua and sometimes helped his students was considered a betrayal of Chern—both by Chern himself and by his even more zealous allies.

I spoke some years earlier, for example, with Sheng Gong, one of Hua's last students and an expert in several complex variables. Gong was then headed to Princeton to meet with Joe Kohn, and friends told me that a Chern stalwart was trying to stir up trouble by claiming that I had enlisted Gong and other Hua partisans in order to mount an attack on Chern. Kohn, as might be expected, dismissed this outlandish-sounding allegation, probably having no idea of the bizarre context in which it was made. I assured Kohn that the whole thing was ridiculous.

More nonsense ensued in 1984, shortly after I arrived in San Diego. My friend Ngaiming Mok, who was still at Princeton, told me about a party he attended with many Chern devotees. Someone mentioned the fight that Chern and I were purportedly waging, and to spice things up, he placed a call to Chern—in the presence of the entire group—asking him leading questions about all the horrible things I had supposedly done. Mok was taken aback by this curious form of entertainment but felt, as a junior faculty member, that he was in no position to put a stop to it.

There's little doubt that my relationship with Chern was hurt by some of his self-professed loyalists, who seemed driven to cast me in an unfavorable light, perhaps thinking that would enhance their standing with "the master." Even so, Chern and I never stopped talking to one another. We continued our correspondence and met with each other sporadically, right up until his death in 2004. But even Chern's death did not put a stop to the mischief wrought by some of his most fervent followers. (I won't call them "supporters," because I don't believe their activities supported Chern and may have achieved the opposite.)

While that tale continued to play out in its tiresome fashion, another story—of a much more exciting and consequential nature—was beginning to unfold. My former assistant at IAS, Gary Horowitz, was a physicist, as I've already mentioned. I have often hired physicists as postdoctoral researchers because it gives me a chance to keep up with their field and simultaneously indulge my fascination with the realm of intersection between physics and mathematics. During Horowitz's two-year stint at IAS, I had several conversations about the Calabi conjecture with him and other physics colleagues then at the Institute, including Andrew Strominger and Edward Witten. My proof, I told them, was motivated by physics, specifically the notion that even in a vacuum, a space with no matter, gravity could still exist. I felt certain that this must be important for physics, though I was not sure of the exact ramifications. But they didn't show much interest, at least initially.

Things changed, however, in 1984, after I'd left IAS. By that time, Strominger and Horowitz had left IAS too, both moving to the University of California, Santa Barbara. Strominger was exploring a hot new idea called string theory with Philip Candelas, a physicist and mathematician then based at the University of Texas. String theory represents a bold attempt to unify the two most successful physical theories of the twentieth century, quantum mechanics and general relativity, which have the unfortunate characteristic of being incompatible with each other. Quantum mechanics is almost perfectly accurate when it comes to describing the behavior of very small objects or particles in settings where gravity is extremely weak. General relativity does just as well in describing large, massive objects in settings where gravity is strong. But neither theory on its own can handle conditions—which might be found, for instance, in a black hole's interior or during the Big Bang—where an enormous mass is compressed to a tiny size. Nor can physicists carry out meaningful calculations by combining the equations of those two theories, for that admixture merely leads to gibberish.

In the early to mid-1980s, a growing cadre of researchers came to believe that string theory could bridge that gap in physics by providing a new, overarching framework, which held that matter and energy, at the smallest, most fundamental level, are composed of tiny, vibrating strings rather than pointlike particles. String theory further postulated that we inhabit a ten-dimensional universe consisting of the three familiar (and infinitely large) spatial dimensions, one dimension of time, and six additional miniature dimensions that are wound up into a tight coil and thereby hidden from view. The question that Candelas and Strominger, among others, were grappling with concerned the geometry of the six shrunken, or "compactified," dimensions. What, exactly, is the shape into which these extra dimensions are confined?

Strominger knew that they needed a manifold, or space, with well-defined properties, including a special kind of symmetry called "supersymmetry," which turns out to be an intrinsic feature of the manifolds, of the variety called Kähler, whose existence I had proved. Supersymmetry is also a requisite feature of many versions of string theory, which is why it's sometimes called "superstring theory" instead.

After conferring with Horowitz—who was, by virtue of our association, more familiar with my work—Strominger called me up to learn more about Calabi-Yau manifolds and how they might fit in with string theory. I was sitting in my wife's office in La Jolla at the time, gazing out at the expanse of the beautiful, blue ocean, which stretched all the way across to China. And at that same moment, I felt the possibilities for these geometric constructs expanding as well—not only blending with physics but also merging with the all-encompassing sea before me and the universe that enveloped it.

I told Strominger that the six-dimensional form of these manifolds could indeed meet the specifications called for by the theory—at least based on the information I'd received so far—which was just what he'd been hoping to hear. Strominger then met with Witten, who had independently arrived at a similar conclusion. Witten even flew out to San Diego to spend a whole day with me, talking about how to construct the new manifolds using the techniques of algebraic geometry.

Shortly thereafter, the four physicists—Candelas, Horowitz, Strominger, and Witten—joined forces and wrote "Vacuum Configurations for Superstrings," which was published in 1985. That landmark paper, considered a part of the "first string revolution," argued that the six extra dimensions of the theory must be curled up into so-called Calabi-Yau manifolds. The precise shape of those manifolds in turn would determine the kinds of particles that exist in nature, their masses, the strength of the forces between those particles, and other physical characteristics. "The code of the cosmos," claimed the physicist Brian Greene, "may well be written in the geometry of the Calabi-Yau shape."

The "vacuum configurations" paper provided an essential bridge be-

A rendering of the three-dimensional cross section of a six-dimensional, "quintic" Calabi-Yau manifold. (Courtesy of Andrew J. Hanson, Indiana University.)

tween the four-dimensional universe that humans can perceive with their own senses and the underlying ten-dimensional universe posited by string theory, much of which is concealed from view by virtue of its miniscule size. With this piece—combined with other recent advances, most notably achieved by the physicists Michael Green and John Schwarz—string theory suddenly became the rage. Hopes were raised that it could provide the kind of unification in physics that Einstein tried to realize, unsuccessfully, during the last thirty years of his life.

I caught a bit of "string fever" as well, not only because of the central role Calabi-Yau manifolds played in the theory's inner workings but also because I was often asked to explain the abstract geometry integral to the theory, which many physicists were unfamiliar with at the time. This led to a huge collaboration between mathematicians and physicists that lasted for many years. I got swept up in the tide, and it was an exhilarating time that is still leading to interesting developments in both physics and math.

Even now, more than thirty years after the term "Calabi-Yau" was

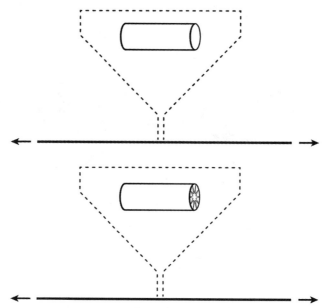

Picture an infinite, four-dimensional space-time as a line that extends endlessly in both directions. A line, by definition, has no thickness. But if we were to look at that line with a magnifying glass, we might discover that it has some thickness after all. This line could, in fact, harbor an extra dimension whose size is set by the diameter of the circle hidden within (*top*). String theory takes this idea of a hidden "extra" dimension and expands it considerably (*bottom*). If we were to take a detailed look at our four-dimensional space-time, we'd see that it's actually harboring six extra dimensions, curled up in the shape of a Calabi-Yau manifold. No matter where you slice this line, you will find a hidden Calabi-Yau manifold, and all the Calabi-Yau's exposed in this fashion would be identical. (Based on original drawings by Xianfeng [David] Gu and Xiaotian [Tim] Yin.)

coined by Candelas and his colleagues, a Google search of that conjunction yields about four hundred thousand results. Furthermore, *Calabi-Yau* is the title of a 2001 play, "Calabi Yau Space" is the name of an album by the Detroit band Dopplereffekt, the Italian artist Francesco Martin has included "Calabi Yau" in the title of several paintings, and in a 2003 story in the *New Yorker* Woody Allen referred to a woman smiling and "curling up into a Calabi-Yau shape." The expression has been used so often, in fact, that I sometimes feel as if my first name is Calabi. That's okay with me, as I admire the man and am proud to be associated with him. Calabi, for his part, has said, "I don't mind that my name and Yau's may be linked forever."

One question that weighed heavily on Strominger and Witten concerned the number of Calabi-Yau manifolds that existed. Strominger put that question to me in 1984, hoping there was just one solution, which would have made life easier for him and others trying to build the theory. At that time, I had only two solutions in hand—two Calabi manifolds that had already been constructed. Yet I soon found many more and, on the basis of those constructions, realized that the ultimate answer would be much bigger. My best estimate was that there were at least ten thousand Calabi-Yau manifolds, each representing a different solution to the equations of string theory and each of a different topological type. We now know that many more Calabi-Yau manifolds can be constructed, going far beyond my initial estimate. I've since posed the conjecture, which is still unsettled, that the number of six-dimensional manifolds (or three-dimensional in complex coordinates) is finite (though still quite large).

But back in 1984, in the early string era, Strominger was dismayed by my response, as things would have been much simpler from a theorist's standpoint if there were just one such manifold or only a handful. I delivered the same disappointing news to a large crowd of physicists attending one of the first major string theory conferences, which was held in March 1985 at Argonne National Laboratory in Illinois.

Many of the top people in the field, and top theorists in general, came to Argonne to speak and present papers, including David Gross and Gerard 't Hooft (two future Nobel Prize winners in physics), as well as the aforementioned Green, Schwarz, and Witten. I, too, presented a paper on the geometry of Calabi-Yau spaces, although it had the more technical-sounding title "Compact Three-Dimensional Kähler Manifolds with Zero Ricci Curvature." (This paper, as you can see, refers to three-dimensional manifolds; however, the Kähler manifolds in question are of three *complex* dimensions, or six *real* dimensions, in keeping with the requirements of string theory.)

Before going to Argonne, Horowitz, Strominger, and Witten had asked me to construct a Calabi-Yau manifold with an Euler number of 6 or –6. The Euler number (or characteristic) is an integer, positive or

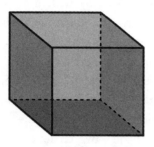 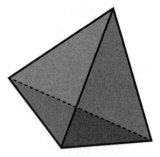

The Euler number, or characteristic, was originally created to classify polyhedra, but it has since been generalized as a way of describing other, more complicated topological spaces (including Calabi-Yau manifolds). For polyhedral surfaces, the Euler characteristic is equal to $F + V - E$, where F is the number of faces, V the number of vertices, and E the number of edges. Using this simple formula, one can see that the Euler characteristic for a cube is $6 + 8 - 12 = 2$. The Euler characteristic for a tetrahedron is $4 + 4 - 6 = 2$. Leonhard Euler was the first person to recognize (in 1750) that all convex, three-dimensional polyhedra have an Euler characteristic of 2.

negative in value, that offers a simple way of classifying topological spaces and showing which ones are equivalent. To take a straightforward example, a tetrahedron, or "triangular pyramid," consisting of four triangular faces, has an Euler number of 2, which can be derived by adding the number of faces (four) and vertices (four) and subtracting the number of edges (six).

My physicist friends wanted a much more convoluted Calabi-Yau manifold with an Euler number of 6 or −6 because Witten had previously demonstrated that the number of families of particles equals half the absolute value of the Euler number. A manifold with an Euler number of 6 or −6 would therefore yield the correct answer, giving rise to the three families of particles that are an indispensable feature of the Standard Model of physics.

And that was one of the paramount goals—to show how string theory could reproduce the Standard Model, the physics we knew, and then carry us far beyond. But the Calabi-Yau manifold that Candelas and company had originally worked with yielded four families of particles, which was close to the desired three but not close enough. Being off by one in this case is a significant discrepancy—like losing track of a child in a

"Home Alone" scenario—constituting an issue that needs to be remedied as quickly as possible.

I didn't have time to get to this problem before leaving for Argonne, but I worked on it during the flight from San Diego to Chicago and came up with a solution—a Calabi-Yau manifold with an Euler number of –6—just before we landed at O'Hare International Airport. I was eager to spread the news but first had to find the person who was supposed to take me to the national lab. A guy approached me at the airport, whom I assumed was my prearranged driver, but when I got into his vehicle I found out that he had never heard of Argonne and had no idea where it was. It should have been a twenty-five-mile drive, but we took a meandering route. He ended up charging me $50 for the fare, and I had to bargain aggressively with him to get it down to that number.

My talk was well received at the conference. I'd been to many scientific assemblies before, but this event stood out because of the air of excitement that infused the proceedings with a sense of optimism. I was also struck by how focused the conference was. People weren't there simply to present their own particular findings. It felt as if there was just one problem at hand, albeit a big one, which everyone wanted to work toward solving. A number of representatives from the media were in attendance too, hanging on every word. Everyone had the sense that this gathering could be momentous—that science might be approaching a critical and long-awaited threshold.

It was not possible to maintain that level of enthusiasm over the long haul, however, as string theory had set an extraordinarily ambitious agenda, the details of which would take years to work out, if indeed they could ever be completely worked out. While some of the earliest, loftiest hopes—such as creating a so-called theory of everything—have yet to be realized, and may never be realized in full, string theory has made many contributions to physics and mathematics that were not anticipated early on. And even though string theory may not be the ultimate theory of nature, it does appear to be at least a step in that direction. And the theory still continues to surprise us in interesting ways. So, in the end, the fact that it might fall short of the original expectations placed upon it should not be viewed as a failure.

For my part, I returned to UCSD energized by the Argonne symposium and eager to work on string theory. I hired a physics postdoc from

Caltech, Brian Hatfield, who was well versed in the subject. I had fifteen graduate students that year, and I showed them the Calabi-Yau manifold I had constructed en route to Chicago, as well as the method I had employed. My student Gang Tian suggested that my approach could be used to construct a few more examples of manifolds with an Euler number of −6, which he then did. It was later discovered that even though these "new" manifolds looked different, they were in fact deformations of the first manifold I constructed—made by bending and stretching without tearing—and hence were topologically equivalent to the original.

I continued to investigate various facets of Calabi-Yau manifolds, owing to their importance both in string theory and in mathematics, and later edited a book called *Mathematical Aspects of String Theory*. But as I've said before, I rarely work on just one thing at a time, so string theory did not consume all my attention. I was still intent on building up geometric analysis—the mathematical discipline out of which the Calabi-Yau manifolds of string theory arose—and my "co-conspirators," Schoen and Hamilton, were working alongside me at UCSD toward that end.

Schoen and I resumed our usual collaborations, such as classifying manifolds with positive (scalar) curvature and talking about the Yamabe problem on compact manifolds, which Schoen was close to solving and did, in fact, solve later that year (1984)—one of his major achievements. We also resumed a lecture course at San Diego that we had started at IAS in 1982 and continued a year later at MSRI. In these lectures, we presented our original work, including ideas that had never been published before. Sometimes we were up until midnight and later, preparing for the next day's lecture.

We needed someone who could take good notes so that we could preserve all of our findings and ultimately present them in the form of two books, *Lectures on Differential Geometry* and *Lectures on Harmonic Maps*, which were published several years later. As I was often trying to help Chinese scholars come to the United States—where they could get a taste of the research environment, as well as earn a decent salary—I asked Yang Lo whether he knew of a qualified person for this job. A researcher named Hsu from the Chinese Academy of Sciences eagerly volunteered. I paid him for more than a year to carry out this work, and that turned out to be a big mistake. Although Hsu was reasonably proficient in mathematics, he was not up to speed in our area. He didn't understand many things, though he never asked Schoen or me for help. He sometimes asked questions of my students, but they didn't treat him well, as he was much older than them, nor did they want to take the time to explain things to him.

In the end, the notes that Hsu compiled were worthless. This was a huge blow to Schoen and me because we often didn't have a chance to write everything down, and much later, after we realized that Hsu had not been up to the task, we weren't always able to reconstruct our arguments in full.

But things got even worse when Hsu was asked to send a progress report to the Chinese Academy. Rather than admit that he had been unable to perform his job, he turned it into a report (or more accurately an attack) on me, claiming that I was plotting against Chern. Hsu further alleged that I wanted to form my own party—"Yau's Party," as he put it—whose entire raison d'être, apparently, would be to stand in opposition to my former advisor. The whole thing was so preposterous that people at the academy understood that Hsu had made it up. Yang Lo was embarrassed; he mailed a copy of Hsu's letter to me and apologized for having sent him to San Diego. Hsu's departure soon followed.

My life was punctuated by moments like this of what seemed to me to be lunacy. Fortunately, as a counterbalance, the mathematical side of things at UCSD had gotten off to a good start. I enjoyed working in close proximity to Schoen, as I always have. Our interests and ways of thinking matched well, making for a successful collaboration—probably the best I've ever had. Michael Freedman, who had recently completed his celebrated proof of the four-dimensional Poincaré conjecture, often joined in our conversations, bringing a fresh and welcome perspective to these exchanges.

I also spoke often with Hamilton, whose office was adjacent to mine, and having him next door was a real treat for me. One topic we discussed concerned some work I had done with Peter Li a year or so earlier called the Li-Yau inequality. Li and I had developed equations related to geometric flow that describe how heat or some other variable propagates over a surface, continuously changing in time. I told Hamilton that the "Li-Yau estimate" could be useful in understanding how singularities—places on a surface, such as pinches or folds, where space compresses to a single point—can develop in Ricci flow and, more importantly, how such singularities can be smoothed over. I persuaded Hamilton that this approach would be critical in tackling the still-unsolved three-dimensional Poincaré conjecture, although the Li-Yau inequality would have to be incorporated into his much more elaborate, and more nonlinear, Ricci flow model.

It took Hamilton a half dozen or so years to complete the latter task— a key step along the path toward an eventual Poincaré proof. A number of my graduate students, including Huai-Dong Cao and Ben Chow, worked with Hamilton on problems involving Ricci flow.

A former classmate of mine was a trustee of a Hong Kong foundation that provided money for a program I ran at UCSD during the summer of 1985 to train mathematicians from mainland China, Hong Kong, and Taiwan in doing research. About forty students, postdocs, and faculty members came to San Diego for this summer session. Hamilton, Schoen, Freedman, I, and others gave lectures, and most participants learned a lot. One of the attendees, Shi-shyr Roan, who had come from National Tsing Hua University in Taiwan, ended up coauthoring a paper with me on constructing Calabi-Yau manifolds using a so-called toric method developed by David Mumford. Roan later worked as my postdoc, and I helped get him a job at the Institute of Mathematics at the Academia Sinica in Taiwan.

Many others who came to the program found it worth their while. Although mathematics was the main attraction, people also found time to play volleyball on the beach—one of the benefits of the San Diego locale that places such as Princeton and Harvard could not match.

I was entertaining a lot of visitors at the time, not just from Asia but also from Europe, the United States, and elsewhere. Pit-Mann Wong—a mathematician at Notre Dame University who graduated from my high school in Hong Kong and also happened to be a native of my hometown, Shantou—came during my first year at UCSD, looking for research suggestions. I showed him an early, handwritten draft of a paper I'd done on the Yang-Mills equations with Karen Uhlenbeck, telling him that after he'd read it, we could talk about certain ideas that might be developed further. (Our paper, which I'm still quite proud of, came out in a 1986 issue of *Communications on Pure and Applied Mathematics*—the same journal, published by the Courant Institute, in which my full proof of the Calabi conjecture had appeared eight years earlier.) I also told Wong, who was on sabbatical, that he could spend the year in San Diego if he liked, and we could delve into other questions involving Yang-Mills equations that I was still curious about. But Wong had already made arrangements to spend his sabbatical at Harvard working with Yum-Tong Siu.

Coincidentally, I soon heard from my former postdoc Jia-Qing Zhong, who was also visiting Siu at Harvard. Zhong called to say that Siu was extremely upset, claiming that Gang Tian had copied a result from a lecture that he (Siu) had presented during a 1985 conference at Columbia. In a letter that Siu sent to me a year or so after the conference, he said, "I do not see how independent your student's derivation can be." Tian's paper, according to Siu's letter, "seems to be somebody claiming as his own a reformulation of a method of mine already presented in public lectures."

To settle the question of originality and priority, Siu asked to see a copy of Tian's full manuscript. I was reluctant to hand it over to him, or to force Tian to do so, because I like to give younger people the benefit of the doubt. On the other hand, I must confess that I wondered whether Tian might have appropriated material from Siu, as charged.

But I was still eager to find a way to resolve this situation. I suggested to Siu that Tian be allowed to give a talk on the subject; other people could then decide whether Tian had copied or not. This, however, did not mollify Siu, who told others—at least as reported to me by colleagues—that I was using my graduate student to attack him, an accusation that had no basis in truth. Siu held on to this grievance for several years, longer than I thought was warranted, but I later saw that his concerns about Tian might have been justified.

Around the same time, I got a letter from Yitang Zhang, a Chinese mathematician who had earned a master's degree at Peking University in 1985. Zhang wanted to pursue work in his field, number theory, at San Diego, so I arranged for him to study with Harold Stark, an outstanding UCSD number theorist who was later elected to the U.S. National Academy of Sciences.

These plans, however, seem to have been thwarted by Shisun Ding, who had become president of Peking University in 1984. Perhaps Ding was still angry at me because I had not made him a member of IAS a couple of years earlier (even though I hadn't been in a position to make such an appointment). I can't claim to know Ding's motivation, but for whatever reason, alternate arrangements were made for Zhang: Instead of pursuing graduate studies with Stark at UCSD, Zhang went to Purdue and became the graduate student of Ding's friend, Tzuong-Tsieng Moh. Zhang resented the fact that he was forced to change his major to algebraic geometry rather than being allowed to pursue the subject he had a passion for, number theory, simply because of the personal relationship between Ding and Moh.

I knew Moh from my first stint at IAS in the early 1970s, and he was not a number theorist. Ding had essentially given Moh a gift in the form of a talented student, Zhang. Such was the power of a university president in China in those days; he could override a student's choice and compel him to work in an entirely different area of mathematics.

Suffice it to say that Zhang's efforts on the Jacobian conjecture did not go well. I believe he had troubles, at least in part, because he was building upon Moh's work, some of which remained unpublished. As a result, Zhang never published his own work on the Jacobian conjecture, including his dissertation, because its validity hinged on unpublished material. And I imagine that's why, after getting his PhD in 1991, Zhang could not get a tenured position in academia for more than twenty years.

Moh, incidentally, worked on the conjecture for many decades and never proved it, nor has anyone else to this day. Zhang's fortunes, however, changed dramatically in 2013 when he shocked the mathematics world with the breakthrough he'd made on a celebrated problem in number theory, the twin prime conjecture, which dated back to the 1800s.

Apart from antics of this sort going on in the background, things at San Diego were proceeding quite well. In June 1985, I won a MacArthur Fellowship, which came as a complete surprise to me, albeit a pleasant one. A *Los Angeles Times* article about the award described my work in differential geometry as "so complex that his own colleagues don't understand it."

That was, I suppose, an upgrade from an article that had come out in the same paper a year earlier to announce my hiring at UCSD. The same reporter had then described my mathematics as being essentially useless—a characterization that probably stemmed from my response to his question about the societal value of my work in geometry. Research in pure mathematics, I replied, has an important impact over the long run but rarely affects life in the short term. For example, I said, "It probably cannot be used to power a garage door." And it was evidently a small step from there to "useless." That said, I was happy to be a MacArthur Fellow and proud to be in the company of the other distinguished winners that year, including Marian Wright Edelman, president of the Children's Defense Fund; the esteemed literary critic Harold Bloom; the scientist and author Jared Diamond; and the dancers and choreographers Merce Cunningham and Paul Taylor. The award also came with some cash, which is always useful. In this case, I put away most of the money for my sons' eventual college educations.

Life in San Diego was good in other ways too. I got to spend time with my kids and take them to places like SeaWorld and the San Diego Zoo, which was fun for all of us and at times made us feel as if we were living the "American dream." We had pleasant weather and almost constant sunshine, with the beach and waves never too far away.

I'd made a new friend in Bill Helton, a fellow mathematician in the UCSD math department. Elsewhere on the university campus, I had a large and mostly eager group of graduate students, plus a great core of colleagues and very real hopes of building upon that. I was told, in fact, that the university would be able to hire fifteen more faculty members, both junior and senior. I was pushing for people whose work I knew and respected, including my friends Leon Simon and Karen Uhlenbeck, as well as Demetrios Christodoulou, an expert in general relativity who would soon make a big splash.

Things were looking up; all systems appeared to be go. Then I ran into the juggernaut known as departmental politics and had to put those dreams on hold. My vision, it turns out, was not shared by others at UCSD. There did not appear to be much enthusiasm for hiring Simon, Uhlenbeck, and Christodoulou, even though I had been given ample encouragement to build up the department. Perhaps I'd been too outspoken in my support of them and in my dismissal of other candidates who struck me as less qualified, thereby alienating some faculty members in the process.

Around this time, I was taken aback by a phone call I received from Freedman, who had recently been appointed to UCSD's Charles Lee Powell Chair of Mathematics. He contacted me late in 1985, while I was attending a conference at Columbia University and staying with Dennis Sullivan, who held (and still holds) the Albert Einstein Chair at the City

University of New York. Freedman wanted to find out whether he would be getting a Fields Medal in the next year; he thought that Sullivan or I would know, but neither of us did. Speaking out of frustration, perhaps, Freedman let me know that his work was more deserving of the prize than mine because his proof of the Poincaré conjecture contained five new ideas, whereas my proof of the Calabi conjecture had just one. As things turned out, Freedman did receive a Fields Medal in August 1986, roughly a year later. The prize was well deserved, in my opinion, although I didn't appreciate the comparisons he drew between his work and mine, which struck me as gratuitous and of debatable accuracy.

I then got caught up in a boondoggle concerning a proposed applied mathematics center at UCSD, which did not have a happy ending. The Mathematics Research Center (MRC) had been based at the University of Wisconsin, Madison, since its inception in 1956, under a collaborative arrangement between the U.S. Army and the university. The building in which the MRC was housed, Sterling Hall, was bombed in 1970 as part of the student uprising against the war in Vietnam and the U.S. military in general. One physicist was killed during the blast and three others were injured, although none of these people was affiliated with the MRC.

By the mid-1980s, the army was looking for a new home for its center. UCSD wanted to host the facility, and I was asked to take part in the bid. I didn't have much of a personal stake in this matter but was willing to help out. However, when it came time to writing a proposal, our effort ran aground because the applied mathematicians in the group hadn't been able to put together a compelling case on paper. It seemed as if I was going to have to write the thing myself, even though I wasn't an applied mathematician—nor was I frequently sought out for my writing skills. For starters, I got some advice from other experts I knew, including Paul Garabedian, an applied mathematician at Courant, and James Glimm from Stony Brook. That riled up the San Diego applied math contingent, who weren't happy that I'd sought outside help. They also weren't happy that I was trying to get pure mathematicians hired rather than lobbying for people in their fields. Ultimately it came down to the fact that they didn't want me involved in *any* affairs relating to applied math within the department.

As an example, my statistician friend Richard Olshen, who was then at UCSD, wanted to recruit a great young statistician, David Donoho, who had just received a PhD from Harvard. I told him that sounded like a good idea. Olshen then told Murray Rosenblatt, UCSD's resident authority in statistics and probability, that I was eager to see Donoho hired. Rosenblatt got mad at me, saying I should stop interfering with personnel decisions regarding statistics.

Many applied math folks had turned against me, protesting the fact that I'd been given a lead role in the effort to bring the army math center to UCSD. In an attempt to alleviate the controversy, Vice Chancellor Harold Ticho, who had previously worked as a particle physicist, decided to hand the project over to John Miles, at the Scripps Institution of Oceanography, which constitutes its own division within UCSD. But Ticho wanted me to continue to lobby for the center on behalf of the university. I refused to do that, telling him that my only interest in this project had been in building up the math department; I wasn't going to devote my time and energy toward building up the math program at Scripps.

My relations at UCSD were clearly starting to fray. Many applied mathematicians opposed the appointments of Simon and Uhlenbeck, arguing that I had already gotten Hamilton and Schoen, and enough was enough. I let Ticho know I wasn't feeling at home within the department and might have to leave the university. Some claimed I said that as a ploy to jack up my salary, but I was actually looking for a way out.

In the midst of the disagreement over future hirings, Ticho arranged for Freedman and me to have lunch with him in the hopes of clearing the air. Ticho asked Freedman whether he was aware of any problem within the department. Freedman said that he didn't see a problem, which made me feel that if there was a problem, it was my own problem, not the department's. As I didn't feel like I was getting much support, by the end of that meeting I concluded there was not much hope for me at San Diego.

I ended up moving far from there—about as far as you can go and still be in the continental United States. Which was a shame because it's a beautiful place with what is touted to be the world's most agreeable climate. And we were close to assembling a strong math department at UCSD. But many of my fellow faculty members didn't want that to happen—or at least not on my terms, which is fair enough. Some were happy with how things were and didn't feel the need to transform UCSD into a world powerhouse in mathematics. Others probably had different visions of the future, as well as different ideas about the best way of building up the department. I was reminded of the old joke: "How many people does it take to change a light bulb?" Just one, the response goes, "but the light bulb has to want to change."

Hamilton stayed on for several more years. He was happy to be in a place where he could pursue his two passions of mathematics and surfing. In 1996, he joined the Columbia faculty (where reasonably good surfing can be found on nearby Long Island, although not up to Southern California standards). Schoen left UCSD in 1987, resuming his position at Stanford.

Fortunately, I'd been able to line up a new landing place. When I'd been at Berkeley in late 1986, I had met with Raoul Bott, who was not only a great mathematician but also a nice guy, as well as someone I've always admired. He let me know that Harvard was about to make me an offer—again. This would be my third job offer from that university, and it wasn't clear there would be a fourth. I told Bott that things at UCSD were becoming difficult for me. He told me that Harvard was eager to have me, but I should take my time and be sure I wasn't making a rash, emotional decision.

In early 1987, I flew to Harvard to meet with Barry Mazur, the math department chair, who was extremely cordial. After our chat, which set me at ease, Mazur introduced me to Michael Spence, the dean of the Faculty of Arts and Sciences, who later won a Nobel Prize in Economic Sciences. Spence was also very friendly, doing everything he could to make me feel welcome. I found out that he was married to a Chinese woman, the granddaughter of a famous Chinese scholar I had learned about in my youth. The salary that Harvard could offer me was lower than what I was getting at UCSD, but the university was able to give me a good deal on a home loan, which helped compensate for the shortfall in salary.

I was definitely tempted, but what sealed the deal was the fact that the MIT Lincoln Laboratory offered Yu-Yun a job in applied physics, where she'd be able to do the kind of work she enjoyed doing most. I said yes to Harvard, joining the faculty in 1987. I've been there for the past thirty-plus years. Not all of it has been blissful, of course, but it's still been a good, long run, giving me reason to believe there might be something to that old adage—"The third time's a charm."

CHAPTER NINE

Harvard Bound

GOING TO HARVARD IS, in at least one respect, different from going to just another school. I arrived in Cambridge in July 1987 at what the university calls "the oldest institution of higher education in the United States," and, hokey as it may sound, I could almost feel the weight of history hanging in the air. Owing to the presence of historic buildings within close view of the math department—structures like Massachusetts Hall, built in 1718, and Harvard Hall, built in 1766—there was no mistaking the fact that I was joining a place steeped in tradition, nearly a century and a half older than the United States itself. I was not well versed in Harvard lore upon entering, though I did my best to educate myself about some of my illustrious predecessors.

The "Colledge" was founded in 1636 upon land bequeathed to the school by a local minister, John Harvard, who also donated upon his death the entirety of his four-hundred-volume library (which has since grown into a university-wide collection of roughly *seventeen million* holdings). Mathematics books did not figure prominently on the shelves of the original library. Nor was mathematics considered an essential part of the school's early curriculum, as arithmetic and geometry, according to the historian Samuel Eliot Morison, were then regarded "as subjects fit for mechanics, rather than men of learning."

Algebra was not taught at Harvard until the 1720s or 1730s, roughly a century after the school's founding. Another century passed before the first bit of original mathematical research was carried out at the college: In 1832, a twenty-three-year-old tutor named Benjamin Peirce published a proof on "perfect numbers"—positive integers, like 6 and 28, that are equal to the sum of their factors ($1 + 2 + 3$ and $1 + 2 + 4 + 7 + 14$). Peirce, however, was not applauded for the achievement, because math faculty members in that era were supposed to focus on teaching and writing textbooks, *not* on proving theorems.

Things changed dramatically in the early 1890s when two mathematicians trained in Europe, William Fogg Osgood and Maxime Bôcher, became Harvard instructors and eventually full professors. Osgood and Bôcher brought a more "modern" perspective to the school, establishing a research culture within the mathematics department that took hold, gaining considerable momentum by the time I arrived on the scene nearly one hundred years later.

During that century, mathematics had undergone major transformations, and brand-new fields—including category theory, the Langlands program, and geometric analysis—had sprung into existence. Physics, meanwhile, witnessed spectacular advances with the advent of quantum mechanics and general relativity early in the 1900s and, much later, the hope of merging those two successful disciplines into the potentially unifying framework offered by string theory. My interest at the time was squarely focused on string theory, and my friend Isadore Singer, whose office at MIT was just two miles away, was also very excited about the subject. He was well connected too, as I've mentioned, and offered to help me get money from the Department of Energy (DOE) so that I could hire some postdocs for research in that area.

Arthur Jaffe, who had just become the Harvard mathematics chair, asked to be written into that proposal, suggesting that we split the money from DOE, if and when it came through. I went along with Jaffe's request.

DOE insisted that Jaffe and I go to Washington, D.C., to make our funding proposals in person. We were given half an hour for our presentation. Jaffe said that he'd take the first fifteen minutes, and I'd have the last fifteen minutes. His talk went longer than planned, leaving me with just five minutes to make my pitch. But we got the funding, and with it I was able to hire some excellent researchers, including the physicist Brian

Greene, who undertook some truly consequential work as my postdoc (see more below).

About a dozen graduate students came with me from San Diego to Boston. Four of them—Jun Li, Wan-Xiong Shi, Gang Tian, and Fangyang Zheng—enrolled at Harvard. I helped get the others placed at nearby schools—Brandeis, MIT, and Northeastern—while I continued to serve as their advisor.

At Harvard, I joined an impressive faculty, and I had tremendous respect for my peers, who included Raoul Bott, Andy Gleason, Dick Gross, Heisuke Hironaka, George Mackey, Barry Mazur, David Mumford, Wilfried Schmid, Shlomo Sternberg, John Tate, Clifford Taubes, and many others. I soon found myself surrounded by a large contingent of students and researchers from China—so many, in fact, that people often assumed that I worked only with Chinese graduate students. But about a third of my PhD students over the years have been non-Chinese, and I've always accepted any student who, in my estimation, is good enough to be at Harvard.

Nevertheless, I did have a sizable number of visitors from China—enough to attract the attention of the CIA, which periodically asked me to report on what all these people were up to. The details I provided—concerning Calabi-Yau manifolds, Ricci flow, Yang-Mills theory, and so forth—were sufficiently boring that after I'd submitted several years of these accounts, the CIA stopped asking me. Agency officials had evidently concluded that no national security issues were at stake and that the realm of geometric analysis did not fall under their purview.

Life was busy, as has been the case for as long as I can remember—almost going back to the days when I answered (grudgingly) to the moniker "Little Mushroom." I had a large number of graduate students to keep occupied as I settled into a new routine at a new university. I finished teaching by 4 p.m. each day so that I could pick up four-year-old Michael at day care and then get six-year-old Isaac at his elementary school in Belmont, where we then lived, one town over from Cambridge. I played with the boys after school and tried to teach them Chinese poems, although those lessons were not a big hit.

I also devoted a lot of attention to my student Tian. He typically came to my house three times a week, working with me a couple of hours each session—a tradition we had started back in San Diego. I trained him hard because I sensed great potential, but my efforts backfired to some extent. I eventually started to wonder whether Tian might be too focused on getting quick results—a tendency that, if borne out, might lead him to take shortcuts. I've also found that some people resent the fact that you've given them extra help. Rather than being grateful, they turn on you and act as if you've done nothing for them, preferring to reinforce the notion that their success was entirely due to their own efforts. It's similar to what can happen when you loan money to a friend, who then wants nothing further to do with you since your presence reminds him or her of the debt.

But back in 1987, Tian and I were still quite close. He got his PhD in 1988, and I wrote a strong letter of recommendation for him. Princeton offered him a position, though I heard from a mathematician in the department that Siu had aired his concerns about Tian. I wasn't trying to defy a Harvard colleague, Siu, by means of my avid support of Tian; I was merely trying to help launch my student in his new career, which is a normal thing for an advisor to do.

Nevertheless, I belatedly realized that Siu might have been on to something and that my confidence in Tian was perhaps misplaced. Several years later, Tian told me that he had figured out a way to solve the so-called Yau conjecture, which would have been a very interesting development. (Tian sometimes referred to this as the "Yau-Tian-Donaldson" conjecture to attach his name to the notion, although Donaldson himself called it the "Yau conjecture" because the idea had originated with me.) In a conversation with Singer around that time, I casually mentioned my former graduate student's accomplishment. Singer—who'd just been named Institute Professor that year, MIT's most prestigious appointment—was extremely influential on the institute's faculty. With Singer's backing, MIT soon offered Tian a faculty position, which he gladly accepted.

By the time Tian joined MIT's math department in 1995, he still had not written up the paper in question. In fact, he did not post a proof of the full conjecture on the electronic archives until September 2015, *twenty years later,* and this paper came out eighteen months after a full proof had already been published, electronically, by Xiuxiong Chen, Simon Donaldson, and Song Sun. Looking back now, with the benefit of two decades of hindsight, I wish I'd been more cautious in my conversations with Singer.

The publication of the aforementioned papers did not put the matter to rest because Tian had claimed—in a talk at Stony Brook on October 25, 2012—to have completed the first full proof of the Yau conjecture, which was a problem that Donaldson and his colleagues had been working on for some time, while making significant progress toward that end. About eleven months after that talk, when Tian still had not furnished his proof, Chen, Donaldson, and Sun went public with their grievances, rebutting Tian's "claims on the grounds of originality, priority and correctness of the mathematical arguments." Tian's talk provided "few details," the group said, and they had seen "no evidence that Tian was in possession of anything approaching a complete proof at the time of his announcement in Stony Brook." The work that Tian had provided, they added, contained "serious gaps and mistakes," and many changes and additions subsequently made by Tian "reproduce ideas and techniques that we had previously introduced in our publicly available work."

Donaldson is an extremely talented mathematician of the highest repute—a true gentleman in the field—and I'm not aware of anyone, including Tian, who has convincingly refuted the charges that he and his colleagues raised.

But we've managed to get far ahead of ourselves, so let us return to late 1988 and early 1989, when I was asked to join a National Science Foundation (NSF) panel that was charged with deciding who should get NSF grants in geometry. I served alongside Robert Bryant, Chuu-Lian Terng, and several other mathematicians. As it turned out, I didn't make many comments throughout this process, partly because of NSF rules that prevent one from evaluating proposals written by one's colleagues or former students and coauthors. A lot of the people I knew fell into those categories, so I had to leave the room when their proposals were being discussed. When I was able to rejoin the conversations, I was surprised by some of the severe criticisms and what I felt were unduly harsh opinions expressed by other panel members at some of the proposals in question.

Some time after our work was concluded, I ran into Terng at the University of California, Irvine, where she was then teaching. She told me that the NSF would never ask me to join this panel again because my comments regarding the candidates were too disparaging. Her statement surprised me, given that I hadn't said much compared with other panel members. On the other hand, I realized I had a reputation for being outspoken, sometimes offending people as a result. "Your mere presence," Terng added, "scared people so much that it kept them from speaking their minds."

While Terng's assertion seemed off base to me, she was right about one thing: NSF did not ask me to serve on that geometry panel again. I drew a few lessons from the experience. First, that people can attribute all sorts of motives to you, if they choose to, and you don't have a lot of say in the matter. And second, that you can sometimes have a big impact, for better or worse, just by sitting in a room—especially if you have the kind of face that some people find inscrutable and perhaps even intimidating.

I applied for U.S. citizenship in 1990. One step in this process involved my taking a test at the Boston Immigration and Naturalization Service (INS) office for which I was not adequately prepared. The guy administering the test challenged me with a bunch of questions. For instance, he asked me, "Do you think the U.S. president can declare war without the consent of Congress?" I told him that Congress has to agree, adding that President Nixon had probably cut some corners on that score. The test-giver disagreed with me on the latter point, affirming that Nixon (whatever his faults) had not cut corners on the war declarations front.

Overall, I did well on some questions but not so well on others. The INS officer poked fun at me for my mistakes—a couple of which were, in fact, laughable—but he still passed me on the spot, and my citizenship was granted shortly after that.

Until then, I'd been stateless for a long time. With this new designation of "American citizen," international travel suddenly became a much simpler proposition for me. But the abrupt change in status left me feeling uneasy. I still had strong emotional ties to China, my place of birth, but no official ties or documentation to that effect. I had even considered becoming a Chinese citizen, although I can't claim to having devoted much time, or concerted thought, to that proposition. When I mentioned the idea to Qi-Keng Lu, Hua's former student, he told me that would be a mistake. Lu didn't offer any further explanation, but I followed his advice and dropped the matter.

Shortly after gaining my new citizenship, Schoen noticed my recently acquired passport while we were traveling together to a conference in Japan. He subsequently nominated me to become a member of the U.S. National Academy of Sciences, and that nomination was upheld by the academy. This was an unexpected "perk" of my new status; Eli Stein, an influential analyst at Princeton, told me I could have been elected to the NAS eight years earlier, right after I received the Fields Medal, had I been a U.S. citizen at that time.

A change occurred at the *Journal of Differential Geometry* between the November 1989 and January 1990 issues that left Phillip Griffiths off the editorial board. The journal was owned by Lehigh University, and hiring decisions were made by the managing editor, Chuan-Chih Hsiung, a mathematics professor there. I heard through Hsiung that Griffiths was unhappy about losing his editorial position and might have held me partly responsible, even though I had nothing to do with the journal's personnel moves. Griffiths was a highly visible figure in the world of mathematics—active in the American Mathematical Society and the International Mathematical Union—and not someone you'd go out of your way to antagonize. But it seems that somehow I had unwittingly managed to do just that.

Also on the agenda in 1990 was an AMS Summer Research Institute on Differential Geometry that I was running with Robert Greene and S. Y. Cheng. The three-week event, which was held at UCLA from July 8 to 28, was the biggest summer institute the AMS had ever run, with 426 registered participants and 270 lectures. We decided to dedicate the conference to the celebration of Chern's seventy-ninth birthday (which was actually his eightieth birthday, according to the Chinese way of accounting, because a newborn there is considered one year old at the time of his or her birth). I proposed the establishment of a medal called the Chern Award—sponsored by the *Journal of Differential Geometry*—and Chern heartily endorsed the idea. But after I made an announcement about the medal, Chern decided to call the whole thing off. I heard that his sudden change of heart came after consultation with his friends, though no explanation was ever offered to me.

The summer institute, of course, still proceeded without Chern's presence or the bestowal of an award in his name. I rented a big apartment next to UCLA, which we used for an impromptu family reunion.

My sons came, as did my mother. We were joined by my older sister Shing-Yue, my brother Stephen and his son, and my younger sisters Shing-Kay and Shing-Ho, as well as their children. Everything was set up for a large and festive family gathering, with lots of math on the side—almost the perfect combination for me. Except that my mother became gravely ill. We took her for a checkup and, after an extensive battery of tests, a malignancy was found. She was admitted to the hospital the following night. During surgery the next morning, the doctor saw that her cancer was so widespread there was no surgical solution to be had.

Over the next couple of weeks, I went back and forth between the hospital and the conference, where I gave an occasional talk and sat in on various workshops. At the request of many attendees, I also delivered a series of lectures on 100 open problems in geometric analysis—a continuation (with some overlap) of the 120 problems I had discussed in 1979 during the "special year" at IAS.

After the conference was over, I spoke with Harvard's math chair, Wilfried Schmid, who has always been a great ally to me in the department. Schmid was kind enough to grant me a leave of absence for the fall semester so that I could take care of my mother during chemotherapy and assist her with the array of medical decisions she would soon face. Meanwhile, friends at the California Institute of Technology—including the mathematician Tom Wolff and the physicists John Schwarz and Kip Thorne—helped me secure a Fairchild Fellowship during the fall. Caltech even offered me a nice house on campus, which I turned down in order to stay at my mother's apartment, where I slept in spartan style on a bamboo floor mat.

For a while, my mother's cancer seemed to be retreating, and she was doing well. So I went back to Harvard in January 1991 to teach my classes, while Shing-Yue stayed with her. However, by the time I finished classes in May, the cancer had returned, so I immediately went back to California to be with her. We met with her doctor, who delivered the gloomy prognosis that not much more could be done. But there was still a major decision before us: "Do you want to take extraordinary measures to save her if something should happen?" he asked. My mother said no, deciding there would be no point in trying to stave off the inevitable, eking out a little more time at the cost of considerable discomfort. Nevertheless, she was adamant about seeing her grandchildren again, and that wish, fortunately, was granted. I also promised to take care of my sisters and brothers after she was gone.

My mother died on June 2, 1991, at the age of seventy, which is young by today's standards, although a Chinese saying claims that "a man seldom lives to be seventy years old." That saying is probably outdated, as life expectancy in the country is now about seventy-six, though she was not able to make it that far.

Fortunately, she was able to thank her friends and relatives for their support and love before she died. And almost her entire immediate family, including her children and grandchildren, made it to her bedside before the final moment came. My mother had been in extreme pain, but she seemed to become more serene after the children arrived. Seeing her sons and daughters, and their sons and daughters, with everyone appearing to be well provided for, put her at ease. Our presence seemed to help her get ready to let go, which she soon did.

We spent a few days arranging for the funeral. My unpleasant uncle who had long ago offered to engage us in the duck-raising business was now living in Oakland, California, but he did not attend the service. His wife made an appearance on his behalf, though she did not express any regrets over my mother's death. "I did not come earlier," she explained, "because it is very depressing to see a dying person." Some etiquette books, I imagine, would frown upon making such a pronouncement at a funeral, but at least she was candid. The rest of us, however, struck a more somber tone. My ten-year-old son Isaac summed up our sentiments in a letter: "Today is a sad, sad day. Laughter has turned to sobs."

We barely had time for grieving before we had to start thinking about the next steps, such as what to do with our mother's remains. Ideally, we would have liked for her to be buried next to my father in Hong Kong, but we weren't sure what would happen if we tried to take my mother's remains there, which was then moving from British to Chinese rule. We also considered bringing his remains to the United States, but the more I thought about it, the more I realized that my father had no connection to this country. He never learned English and never wanted to come here. So we ended up purchasing a small plot in a Los Angeles cemetery, burying my mother's remains in an area where many other Chinese people had been laid to rest. Some of our older relatives told us that you should not bury your parents too fast; you're supposed to wait a couple of weeks. We didn't know about that arcane rule, and by the time we found out, it was too late anyway.

It was not until all this business was taken care of and things quieted down that I felt the loss of my mother most acutely. I was again wracked with deep sadness, similar to what I'd experienced in the wake of my father's death, though different this time because both my parents were gone. There was no longer an older generation within the family to defer to; it was now solely upon us to take charge. That was a sobering realization, even if it wasn't going to change my day-to-day life in an appreciable way.

But I also took stock of my mother's final years, having deep regrets over the fact that she had worked so hard, taking care of us for most of her life. She gave up almost everything for her family, doing very little to attend to her own needs and happiness. My poor brother Shing-Yuk had needed almost constant attention up until his death not too many years before. I wish my mother had had more time to relax in her old age—to play with her grandchildren, work in her garden, or do anything else that might have brought her peace of mind. Her life was cut short, and she barely had time to rest.

My mother had been a traditional Chinese parent in the sense that she cared more about her sons than her daughters, owing to the conviction that a family's legacy lies in the hands of its sons. She often told me that she saw my success as her own success, which was part of her extraordinarily selfless philosophy, which was in turn the product of her values and upbringing. While I felt guilty pouring so much time and energy into my career, I knew that it would bring her more pleasure if I did well and accomplished something in the world than if I never distinguished myself. An acute appreciation of the sacrifices my parents made provided me with lasting motivation to throw myself into my work and try to achieve excellence. I didn't need much of a push in that direction, because—despite some goofing off in my early years—I became quite driven as an adolescent after my father's death. And I've managed to keep up a rather respectable pace ever since.

Some exciting work was, in fact, unfolding at Harvard at the time, with much of the initiative coming from my postdoc Brian Greene. And

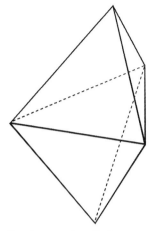 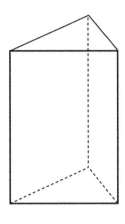

Simple examples of mirror manifolds: the double-tetrahedron (*left*), which has five vertices and six faces, and the triangular prism (*right*), which has six vertices and five faces. These rather familiar-looking polyhedra can be used to construct a Calabi-Yau manifold and its mirror pair, with the number of vertices and faces of the constituent polyhedra relating to the internal structure of the associated Calabi-Yau. (Based on original drawings by Xianfeng [David] Gu and Xiaotian [Tim] Yin.)

while my involvement in the early going was rather minimal, this soon grew into a major research avenue for me as well as many others.

Shortly after Greene arrived at Harvard, he teamed up with Ronen Plesser, who was then a graduate student of Harvard physicist Cumrun Vafa. Building on prior work by Vafa and other physicists—including Lance Dixon, Doron Gepner, Wolfgang Lerche, and Nicholas Warner—Greene and Plesser started toying around with six-dimensional Calabi-Yau manifolds, which were thought to serve as the shape of the "extra" spatial dimensions in string theory. The duo took one Calabi-Yau shape and rotated it in a very special way, producing a mirror image of sorts—albeit one with a very different shape. They discovered that these two distinct Calabi-Yau shapes shared a hidden kinship, both giving rise to the same physics. Greene and Plesser called this phenomenon "mirror symmetry," publishing a paper on the subject in 1990. The two Calabi-Yau shapes that yielded identical physics were called "mirror manifolds."

Mirror symmetry is an example of a "duality"—a phenomenon that comes up rather often in string theory, and in physics more broadly, whereby the same underlying physical situation can be described by two pictures or models that are so different on the surface that they appear to have nothing in common. This idea resonated with me personally because it tied into the notion of yin and yang from ancient Chinese philosophy, and specifically Taoist thought, which stresses the complementarity —and oneness—of two seemingly opposing forces. The concept of duality has led to some remarkable insights in string theory and beyond. Mirror symmetry has been especially productive in this regard.

About a year after Greene and Plesser's breakthrough, the physicist Philip Candelas of the University of Texas and three collaborators—Paul Green, Xenia de la Ossa, and Linda Parkes—performed an extensive calculation designed to test the concept of mirror symmetry. In the course of this work, Candelas and his colleagues exploited mirror symmetry to solve a century-old problem in "enumerative geometry." This is a branch of mathematics devoted to counting objects on a geometric space or surface. The problem Candelas and his colleagues specifically took on involves counting the number of curves that can fit onto a so-called quintic threefold—nonsingular versions of which (i.e., those that do not have any holes) comprise probably the simplest six-dimensional Calabi-Yau manifold to be found. The word "quintic" reflects the fact that this space is defined by a polynomial equation of degree five (including terms such as x^5 or y^5). It's called a "threefold" because it's a manifold of three complex— and hence six real—dimensions.

This problem is sometimes referred to as the Schubert problem because in the late 1800s, the German mathematician Hermann Schubert solved its simplest form, counting the number of curves of degree one (or lines) on the quintic. In 1986, the mathematician Sheldon Katz solved a more difficult version of the problem, involving the number of curves of degree two (such as circles) on the quintic. Candelas and colleagues tackled the next order problem, determining the number of curves of degree three (or spheres) that can fit into a quintic.

And here's how mirror symmetry helped with that task: While the third degree problem was very difficult to solve on the actual quintic, it was much more readily solved on the quintic's mirror manifold—an object that Greene and Plesser had already constructed. Mirror symmetry, Greene explained, offered a way of "cleverly reorganizing the calculation

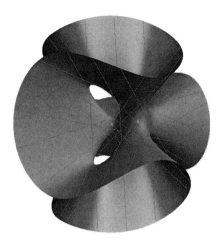

This figure illustrates the general notion of counting curves, or lines, on a surface—albeit on a different surface than those discussed so far. In a celebrated result of nineteenth-century enumerative geometry, the mathematicians Arthur Cayley and George Salmon proved that there are exactly twenty-seven straight lines on a so-called cubic surface, as shown here. Hermann Schubert later generalized this result, which is known as the Cayley-Salmon theorem. (Image courtesy of Richard Palais and the 3D-XplorMath Consortium.)

... to make it substantially easier to accomplish." By doing their calculation on the mirror partner instead of on the original quintic, Candelas's team was able to obtain a precise answer for the number of curves of degree three: 317,206,375.

That certainly got my attention because if their answer was correct, it meant that mirror symmetry might be successfully applied to other problems in enumerative geometry—something that eventually came to pass. Meanwhile, understanding this new concept soon became a top priority for me.

Around the same time, Singer asked whether I would help him run a conference at MSRI on the topic of mathematical physics. Singer's original idea was to focus on "gauge theory," which is closely tied to quantum field theory and particle physics, but I suggested a shift in focus, owing to the exciting new developments in mirror symmetry. Singer was somewhat familiar with the subject, having recently attended a talk Brian Greene had given at Harvard. I told him a bit more, and Singer agreed to hold a weeklong session at MSRI on mirror symmetry in May 1991, asking me to serve as chair.

The meeting turned out to be a charged affair because the early work on mirror symmetry—by people like Greene, Plesser, and Candelas—had been carried out by physicists, and mathematicians didn't yet trust those results and were reluctant to apply them to their own fields like enumerative and algebraic geometry. Such hesitance stems from the fact that deep down, most mathematicians believe they are more rigorous than physicists.

There was already tension in the air at the MSRI meeting when two Norwegian mathematicians, Geir Ellingsrud and Stein Arild Strømme, announced a different result for the third degree Schubert problem, 2,682,549,425, which they had obtained through more conventional mathematical techniques. No one could say for certain as to which answer, if either of them, was correct, but Candelas, Greene, and other mirror symmetry partisans were definitely worried. I went over the calculations with them to see whether anything had gone awry, but we weren't able to uncover any mistakes. Within about a month, however, Ellingsrud and Strømme found an error in their own calculations. They reworked the numbers, this time getting the same answer as the Candelas team—317,206,375—which provided a strong vote of confidence not only for the notion of mirror symmetry, but for string theory as well.

Candelas's work had an even broader reach, for he and his colleagues had produced a general formula for solving the quintic threefold problem for not just lines, circles, and spheres but for curves of *any and all degrees*. That was a bold and sweeping proposition—which had indeed checked out in the case of degrees one, two, and three—but it was still an assertion rather than a proof. In late 1994, Maxim Kontsevich converted this assertion into a precise mathematical statement, which he called the "mirror conjecture."

Shortly thereafter, I started thinking about proving a version of that conjecture, formulated in somewhat different language. After discussing the problem with my former postdoc Bong Lian and my former PhD student Kefeng Liu, we decided to give it a go. While it was an interesting problem in its own right, I was also motivated by the sense that such a proof could provide mathematical validation for mirror symmetry as a whole.

Our forays in this direction soon ran into a bit of controversy. In a paper posted on the math archives in March 1996, the Berkeley geometer Alexander Givental offered a proof of the mirror conjecture. Lian, Liu, and I went through this paper very carefully, and we were not alone in finding it difficult to follow. That raised questions in our minds as to whether the argument was correct. Those concerns were shared by some mathematicians we spoke with at the time, though others seemed comfortable with Givental's work.

My colleagues and I asked Givental to clarify some of the steps we found most confusing, but we were still unable to reconstruct the entire argument. So we decided to start afresh, pursuing an independent proof of the mirror conjecture that was published a year later. Some observers called Givental's paper the first complete proof of the conjecture; others called ours the first complete proof. In an attempt to put the issue to rest, we suggested that the two papers should *together* constitute proof of the conjecture.

People were certainly free to debate this matter further (and some did), but I was ready to move on, for there was an even bigger question at stake—and a deeper puzzle yet to be solved. The proof of the mirror conjecture put Candelas's formula on secure footing, showing that the number of curves on the quintic of varying degrees was not random but instead was part of an exquisite mathematical structure—predicated on a phenomenon, mirror symmetry, that had been discovered by physicists. Proof of that conjecture was indeed a significant milestone, providing an independent check that the intuition coming from physics was justified, yet it did little to explain the phenomenon of mirror symmetry itself. That was something I'd already been trying to do, although proceeding on a parallel track.

It started with a conversation I had with Edward Witten at a 1995 mirror symmetry conference in Trieste, Italy, organized by Cumrun Vafa and others. Witten told me about the new "brane theory" that he was developing with Joe Polchinski and others. Branes were special kinds of surfaces of various dimensions—supersymmetric, minimal submanifolds—that were gaining great importance in string theory and other realms of theoretical physics. One reason physicists got interested in branes is that they greatly generalized string theory. A one-dimensional brane, or "one-brane," is the same as a string, but the theory now had other fundamental ingredients: A two-brane is like a membrane or sheet, a three-brane is like three-dimensional space, and so forth. Thus, researchers had many more building blocks to play with, and the theory became far richer as a result.

Witten discussed some new ideas that the physicists Andy Strominger, Katrin Becker, and Melanie Becker had come up with related to branes, asking me whether these ideas made sense and were natural from the point of view of geometry. I told him that they were natural, realizing a short while afterwards that the mathematicians F. Reese Harvey and Blaine Lawson had essentially hit upon the same ideas earlier, although they had called the objects in question "special Lagrangian cycles" rather than branes.

I started thinking about how these submanifolds, or cycles, might relate to the internal structure of Calabi-Yau manifolds in string theory. My postdoc Eric Zaslow and I began working on this shortly after my return to Harvard. We made progress in particular on the question of what the submanifold of a Calabi-Yau manifold would correspond to in the Calabi-Yau's mirror manifold. We showed, for instance, that a three-dimensional torus, or "donut," mapped (or corresponded) to a point in the mirror.

Strominger soon came to Harvard to interview for a possible job in the physics department, which he eventually got. The three of us joined forces in an attempt to provide a simple, geometric picture of mirror symmetry. The main premise of the resultant SYZ (Strominger-Yau-Zaslow) conjecture, which emerged from our joint effort, is to show how mirror symmetry comes about and how to create mirror manifolds. The basic approach we came up with is to take a six-dimensional Calabi-Yau manifold and break it down into two three-dimensional submanifolds that are modified in a specific way and then put back together again. At the end of this procedure, if done correctly, one will end up with the mirror manifold of the original Calabi-Yau. The methodology advanced by Strominger, Zaslow, and me helps to illuminate the subtle geometric connection between each mirror pair, thereby offering clues as to how mirror symmetry works. Many people, upon reading our 1996 paper, were surprised by the simplicity of our approach.

Thanks to SYZ, Strominger noted, "mirror symmetry was demystified a little bit. Mathematicians liked it because it provided a geometric picture of where mirror symmetry comes from, and they can use that picture without reference to string theory."

Two decades after its inception, the SYZ "conjecture"—which has been proved only in special cases and not yet in a general way—has showed remarkable staying power. It remains an active area of study. And if you can believe the University of Michigan mathematician Lizhen Ji, my for-

The SYZ conjecture—named after its authors, Andrew Strominger, Shing-Tung Yau, and Eric Zaslow—offers a way of breaking up a complicated space such as a Calabi-Yau manifold into its constituent parts, which are called "submanifolds." Although we cannot draw a six-dimensional Calabi-Yau, we can draw the only two (real)–dimensional Calabi-Yau, a torus or donut (with a flat metric). The submanifolds that make up the donut are circles, and all of these circles are arranged by a so-called auxiliary space B, which is itself a circle. Each point on B corresponds to a different, smaller circle, and the entire manifold—or donut—consists of the union of those circles. (Based on original drawing by Xianfeng [David] Gu and Xiaotian [Tim] Yin.)

mer PhD student, the conjecture has served as a "guiding principle for a whole generation of people working on mirror symmetry." Another former student of mine, Conan Leung, continues to turn out fascinating papers on SYZ. Multiple workshops devoted to SYZ and a related topic, "homological mirror symmetry," are being held each year through a collaboration supported by the Simons Foundation (which was started by Jim Simons), involving an impressive group of players from Harvard, Berkeley, Brandeis, Columbia, Stony Brook, the University of Pennsylvania, the University of Miami, and IHES.

Over the past several years, says my colleague Bong Lian, "the geometric and algebraic pictures of mirror symmetry have started to converge. Progress has been made towards encapsulating this idea [mirror symmetry] within one (albeit complicated) formula."

Mirror symmetry has had a dramatic, and surprisingly large, influence on enumerative geometry, algebraic geometry, and many other branches of mathematics. Mathematical conferences on mirror symmetry and SYZ are still regularly held all over the world. It's gratifying to think that this vibrant sector within the math world is an offshoot of string theory and work that was to a large extent originally carried out by my former postdoc Greene and his collaborator Plesser in the late 1980s. Although string theory has not yet proved to be the "theory of everything" that some had hoped for, it has shown its usefulness in mathematics and in many areas of physics. And research in those directions is currently expanding rather than narrowing, which is exciting to consider—as well as to be part of.

Strominger joined the Harvard physics faculty in 1997, and I turned my attention to a series of equations he had drawn up a decade earlier which pertain to more general solutions to string theory that are not limited to Calabi-Yau manifolds. Calabi-Yau manifolds are classified as Kähler, meaning that they are endowed with an internal form of symmetry. Strominger's equations applied to non-Kähler manifolds, which were largely mysterious. That was part of the appeal for me—to explore something new: A lot of tools had been developed in algebraic geometry for studying Kähler manifolds, but not much methodology was available for dealing with their non-Kähler counterparts, which largely resided in terra incognita.

Another reason I was eager to pursue this work was that mathematics offers one of the best checks we have on string theory. Even though we have not yet devised any definitive experiments to test the theory—and that turns out to be extraordinarily hard to do, owing to the fantastically high energy regimes and vanishingly small distance scales involved—we can at least see whether it's mathematically consistent. The general approach is to assume something is right and then work out the mathematical consequences. If the consequences you arrive at make sense, then you know the assumption you started with is at least plausible. We still need to see something in nature to know for sure—empirical validation, in other words—but mathematics can provide the first indication that you're on the right track. And so far, string theory has met the test of mathematical consistency.

Strominger's equations were difficult to work with, but after wrestling with them for many years I finally found some solutions while working with my former PhD student Jun Li, who was then (and remains) a Stanford professor, and later with Jixiang Fu, a former Harvard postdoc who now teaches at Fudan University in Shanghai.

The work with Fu took many years to bear fruit, but his patience and persistence were finally rewarded. When Chinese scholars come to the United States as postdoctoral researchers, they normally want to rack up as many publications as possible. (This sentiment, I should point out, is not limited to Chinese scholars, as the "publish-or-perish" mentality is pervasive throughout academia—often to the detriment of ambitious and risky undertakings.) Fu and I worked for two years before discovering a mistake in our work. He returned to China with nothing to show for his rather considerable labors; but he came back to Harvard again, and this time we succeeded. He eventually had several important papers in his name and was later asked to give a speech at the International Congress of Mathematicians in Hyderabad, India—all of which helped his career prospects. I am impressed that he stuck it out and am also grateful for his forbearance.

That said, this research is still at an early stage because so far my colleagues and I have managed to solve only special cases of the Strominger equations. My friend Melanie Becker, a string theorist at Texas A&M University, told me that if I were to succeed in my broader goal—solving the Strominger equations in their full generality—that would be an even greater accomplishment than the proof of the Calabi conjecture. Of course, success in this venture, by me or anyone else, is by no means guaranteed. Furthermore, it took a long time to figure out the importance of the Calabi conjecture to mathematics and physics. It could take even longer to determine the full implications of the work on Strominger's equations, if indeed they are ever realized.

In December 1997, I went to Washington, D.C., with Yu-Yun and our sons to receive a National Medal of Science. Winning a scientific award, as I've said, doesn't really affect my work, nor does it motivate my research agenda in any way, but this particular outing stood out because we got to meet the president of the United States and attend a party in the White House. The most famous winner among our group was James Watson, a codiscoverer of the double helical structure of DNA. My sons had studied biology in school and were excited that Watson would be there. We'd all read *The Double Helix,* and I liked the book because Watson seemed so honest, although I didn't appreciate the fact that he seemed so comfortable with the idea that he and Francis Crick had taken some of the credit that should rightfully have gone to Rosalind Franklin. That was nothing to be proud of, even if the work, overall, was monumental.

During the White House party, I met Robert Weinberg, a well-known cancer researcher from MIT, and my wife and I chatted briefly with him and his wife. He asked me what I thought about math education, and I told him that I thought math education was very important, though the field suffered because, as I said, "Most of the people in it study only math education; they don't study math." Weinberg then replied, "Professor Yau, my wife is in math education." At that point our conversation got a bit strained.

Before President Bill Clinton came out and spoke, Vice President Al Gore handed each of the winners a certificate. I mentioned something to Gore about his being a Harvard graduate and my being a Harvard professor, but he must not have heard what I said, or perhaps he could not understand my words, as he made no reply. We then waited a long time for Clinton to arrive. Some folks were getting impatient, but when he finally appeared, he was extremely charming and needed to say no more than a sentence or two to make everyone happy. I guess that's what they call charisma, and Clinton—despite some apparent lapses in judgment and behavior—seemed to have no shortage of that.

This award was different from others I'd won, such as the Fields Medal, which almost no one outside of mathematics knows about. The National Medal of Science, by contrast, makes a bit of news. My sons used to think I was the most boring person on Earth and that the work I did couldn't have been more tedious, but they were impressed to see me in the company of celebrities they'd actually heard of, let alone the president of the United States. "Ba always acts like he's smart," my son Isaac had said, although up until then he hadn't seen much evidence of that. But he now had to reassess and told his brother Michael, "Maybe he is actually good."

My neighbors in Belmont, who hadn't given me a moment's notice before, suddenly learned something about me through local news stories that reported on the prize. I wasn't just some Chinese guy who was hard to understand and generally kept to himself. I might in fact be someone worth paying attention to. Like many people from China, I never fit well into the U.S. suburban scene. I didn't play tennis or golf; I didn't coach soccer or Little League baseball; and I didn't find many opportunities to interact with my neighbors. Although I lived next to them, I also lived a world apart. This award didn't change that, but at least some of the people around me had a better sense of who I was and realized for the first time that I was at least somewhat accomplished.

My sons often complained that they, like me, didn't fit in well either, though I tried hard to get them involved in all kinds of "normal" American things. I took them to places like Disney World, SeaWorld, and the San Diego Zoo. I took them to movies at the Fresh Pond Cinema in Cambridge and to the local video store where they rented scores of popular films. I also carted them all over town so they could practice, and compete in, swimming, soccer, basketball, and other sports. I even took them skiing; I would sit in a drafty lodge all day long doing mathematics while they hit the slopes. Michael once complained that no kids came to our house because we didn't have any fun games. So we went to the store together and spent hundreds of dollars on foosball, table hockey, and other ostensible forms of amusement, but that still didn't draw many kids to our house.

Nevertheless, my sons garnered some attention in high school through their prowess in science. Harvard biologist and immunologist Jack Strominger, my friend Andy's father, allowed both of my boys to work in his laboratory, which was something he did not normally do for high school students. I didn't push them toward mathematics because I felt that might put too much pressure on them, given that I was by then a known commodity in the field. The MIT algebraic geometer Michael Artin was often compared to his famous father, Emil Artin, and the same thing happened to the Harvard mathematician Garrett Birkhoff, whose father, George David Birkhoff, was one of the most influential mathematicians of his generation. That's not always the easiest situation to be in, especially for someone starting out in a career.

I thought my boys might benefit from trying their hand in another branch of science, and they both seemed to have an affinity for biology. Isaac and Michael parlayed research projects they undertook in Strominger's lab into entries in the nationwide Intel Science Talent Search. Isaac was a semifinalist in the competition.

Three years later, Michael also took a job in Strominger's lab—after first working at a Gap clothing store in downtown Belmont and discovering that full-time employment can be taxing (although his wardrobe benefited). Michael was initially charged with cleaning Strominger's lab, but he persuaded a postdoc with whom he'd become friendly to help him carry out some experiments on the side. His research project turned out well, and he refashioned it into an entry for the Intel competition, where he became a finalist. That triumph in turn instantly boosted his popularity at high school, and even some girls started to take an interest.

I was grateful to receive a letter from the Talent Search saying that Michael had named me as "the one person who has been most influential in the development of my scientific career." Although my father was not a mathematician, he was the one who inspired me most of all to become a mathematician, and my mother was the person who did more than anyone else to support me until I could realize that goal. I was glad that I'd been able to fulfill a similar role for my sons. After majoring in biology at Harvard, Michael went on to earn an MD from Stanford Medical School. Isaac got a PhD from Harvard Medical School, where he now teaches microbiology and immunology.

While "honor thy father" is a basic dictum in Chinese culture, I'm acutely aware of the fact that my wife, who's had a long career in physics, has had an equally big role, if not bigger, in teaching our children and cultivating their interest in science. It's truly been a team effort in our case, and Yu-Yun and I were pleased that our sons fared well in the American educational system.

At the same time, we had concerns that we might have gone too far in trying to afford them a lifestyle that was similar to that of their peers in terms of recreational activities, entertainment, and the like. The process of "Americanization" had apparently come at a price, as they seemed to be losing their awareness and appreciation of their Chinese heritage. Led by Michael, the boys started to resist their Chinese language studies. What was needed, we realized, was a countervailing strategy to offset this disturbing trend and help reacquaint them with their native roots.

CHAPTER TEN

Getting Centered

WHEN ISAAC GRADUATED from high school in 1998, we took both him and Michael on vacation to China. Rather than the standard tour, sticking to big and somewhat modernized cities like Beijing and Shanghai, we headed to more rugged terrain: Xinjiang Province in far northwestern China, a remote and hauntingly beautiful region wedged between Tibet and Mongolia. After hiking to spectacular alpine lakes that rival the splendor of Banff's Lake Louise (minus the hordes of tourists) and taking in other natural attractions, we flew to the city of Dunhuang in the neighboring Gansu Province.

Dunhuang is famous for the Mogao Caves, also called the Caves of the Thousand Buddhas, burrowed into cliffs just south of town. Tens of thousands of ancient scrolls, tapestries, and other relics dating back to 400 CE have been recovered from these caves, and in some cases pilfered, carted off by the truckloads to museums throughout the world. Dunhuang also sits on the edge of the Gobi Desert, and we spent three extraordinary days driving across it—the largest desert in Asia—stopping at oases along the way, before reaching Lanzhou, an ancient city that once served as a major link on the historic trade route known as the Silk Road.

We had taken the boys to China periodically to keep them connected to their roots. In 1991, we had had a much more extended visit to Asia, partly in response to a rhetorical (and mildly annoying) question Michael had posed: "What do I have to learn Chinese for?" We never answered his question directly, but we did respond to it in a forceful manner. I took a sabbatical for the 1991–1992 academic year, lining up a visiting professorship at the National Tsing Hua University in Hsinchu City, Taiwan. We arranged for our boys to spend an entire school year in Hsinchu where Yu-Yun and I hoped they'd master Mandarin while also gaining in-depth exposure to the culture.

Ironically, traditional Chinese culture was much stronger in Taiwan than on the mainland because, as a key element of the Cultural Revolution, Chairman Mao had called on his followers to "Smash the Four Olds" and thereby lay waste to "old customs, old culture, old habits, and old ideas." With regard to old ideas, members of the armed revolutionary youth corps known as the Red Guards even attacked and vandalized the tomb of Confucius. That move had great symbolic significance, indicating that ancient belief systems were upended during the Cultural Revolution, leaving behind somewhat of a spiritual and philosophical void in Chinese life.

Michael and Isaac would have been in second and fifth grades had we stayed in the United States that year. However, because of their sub-par performance on the Chinese language part of the entrance exam, we were told they had to enroll in first and second grades in Taiwan. Yu-Yun and I insisted that they enroll in second and fifth grades so they could be with children their own age. Chao-shiuan Liu, the president of Tsing Hua University (and the eventual premier of Taiwan), backed us up, and our wish was granted. It was a big challenge for our boys because they were at the first-grade level in terms of their verbal and written skills in Mandarin, so they had to work extra hard to keep up. During our first two months, Yu-Yun spent a couple of hours with them each day to help them learn the language.

Peer pressure helped as well. I've found that when you give kids a strong incentive to learn a language—and in this situation, they didn't have much choice—they can pick it up quickly. In fact, our sons caught up with their classmates during the course of the school year. Not only that, they started winning swimming competitions too, which came as a complete surprise to me, given that I had to drag them to the pool and force them into the water to get them to swim back in the United States.

While things went well for my sons, the climate within Tsing Hua's math department was not ideal for me. Because I had an international reputation, some members of the faculty were worried that I would become too influential and take over the whole place. A few of the department's older professors decided not to attend my lectures as a sign of disrespect. Others tried to find a different way of undermining me. Since they couldn't challenge me on the grounds of pure mathematics, they argued that with me in the department, pure mathematics would siphon off too many of the school's limited resources, and applied math would suffer as a result. That was the same reasoning that some folks had employed at San Diego. The main difference was that Tsing Hua president Liu ignored their arguments. I had his full support and did not get overly concerned, knowing I'd be there for just a year.

One nice outcome from my stay at Tsing Hua was that it brought me into contact with three excellent students—Ai-Ko Liu, Chin-Lung (Dragon) Wang, and Mu-Tao Wang—who traveled more than an hour to the university to hear my lectures. These students showed a great interest in math and later ended up earning PhDs at Harvard under my supervision. I still collaborate on general relativity problems with Mu-Tao Wang, who teaches at Columbia, and maintain occasional contact with my other former students.

During my 1991–1992 stay in Taiwan, I also made frequent visits to my alma mater, the Chinese University of Hong Kong. The physicist C. N. Yang, who joined the CUHK faculty in 1986, informed me of his plans to build a mathematics institute in Beijing when we met in 1992. Yang felt that for China to catch up with the rest of the world in science, the fastest route would be through mathematics, where one wouldn't need so many resources, such as what physics or biology require. He was confident that he could secure funding of $1 million per year from the Chinese government and an additional $1 million per year from private sources. He asked me what I thought about all this and what I could do to help.

I told Yang his plan sounded good though it would make sense to run the idea by Chern first, for in 1984, Chern had established a mathematics research institute at Nankai University in Tianjin, China. We should clear things with Chern, I said, before moving ahead with plans for a new center. I'm not sure what Chern said to him when the two of them eventually spoke, but Yang soon dropped the whole proposition and his attitude toward me seemed to change drastically.

Something else changed as a result of my morning meeting with Yang. I had casually mentioned to him that I would be having lunch that day with a wealthy Hong Kong businessman, Jimin Cha, who was friendly with both Yang and Chern. I hadn't told anyone else about my meeting with Cha, which concerned fundraising for a newly proposed math center at CUHK. However, when I got together with Cha, I was told just beforehand that we would not be talking about fundraising at all, which makes me think that someone might have intervened. I don't know who that "someone" was, but the list of suspects in this case seemed rather small.

Before my discussion with Cha, I had been asked to start a mathematics center by CUHK president Charles Kao (who later won a Nobel Prize in Physics for developing the use of fiber optics in telecommunications). I decided to go ahead with this proposition in part because it touched on something my father had said when I was young—that, being Chinese, I should do something for China someday. This could be a way of advancing that goal—of "giving back," so to speak—as there was a major shortage of well-trained people in the field of math. A research culture in mathematics, moreover, was seriously lacking. I believed that China would never become a truly modern society, making use of the latest in technology, unless its general public was conversant in math and its educational system turned out a fair portion of the field's leaders.

Over time, a large group of talented people from throughout the region could be trained at a facility like this. By starting such a center—and a series of centers, as things eventually turned out—my ultimate hope was that one day China's achievements in mathematics would be comparable to those of the United States and Europe. My involvement in this endeavor also offered me a way to "find my center," to achieve a balance between my career in the West and my roots in the East. And I've been at it ever since, spending most of my summers in Asia after Harvard's school year ends. This effort has not only helped me realize my goals of training mathematicians in China, Hong Kong, and Taiwan, but it has also been a source of satisfaction for me—personally, spiritually, and psychically—as well as a source of frustration, which will later become clear.

In this instance, I didn't have to worry about gaining Chern's approval, because he didn't care about CUHK and would not view anything going on in Hong Kong as a threat. His focus had always been on mainland China. Yang, on the other hand, was much more invested in CUHK. And he probably would have been much more supportive of the new mathematics center had Kao assigned the leading role to him rather than to me.

Chern's institute in Nankai was completely funded by the Chinese government, but I didn't want to rely on government financing, given that the national economy was still quite weak. Kao said he could get about $2 million from the university to start things off. I thought I'd try to raise the rest of the money through private donations, and perhaps the best place to do that in Hong Kong would be at the Jockey Club—"a prestigious institution," according to the *Wall Street Journal,* that "stands apart as a money-making machine." Established in 1884, the Jockey Club is a nonprofit organization that was long ago granted a monopoly on horse racing and lotteries. A lot of gambling goes on in the territory, and the proceeds have rolled in ever since the club opened its doors (and cash registers), making it Hong Kong's largest single taxpayer as well as its largest community benefactor.

The Jockey Club provided a substantial amount of the funding for the establishment of the Hong Kong University of Science and Technology (HKUST), which opened in 1991 on a 150-acre campus that overlooks Clear Water Bay. I didn't go to the club to bet on horses; I went there to bet on math, hoping to meet some high rollers who might be willing to donate a relatively minor sum for a new math center as opposed to the roughly half billion U.S. dollars that had been needed to start HKUST. Getting businesspeople to contribute money for medical causes is not always difficult, but persuading them to support mathematics—which may seem removed from peoples' everyday lives—can be a much harder sell. It sometimes helps when I explain how crucial mathematics is to engineering and practically all branches of science, including computer science.

As luck would have it, one of the first people I met as part of this fundraising effort was William Benter, who already had a deep appreciation of mathematics. Although we did not meet at the Jockey Club—a place to which Benter had no formal ties—he was quite active in Hong Kong gaming activities, where he had done quite well for himself. Benter had in fact acquired a fortune through gambling, partly owing to his expertise in computers. Born and raised in the United States, he started his gambling career at the blackjack tables of Las Vegas. In 1984, he moved to Hong Kong, where he developed computer software to calculate the outcome of horse races. Before long, he was earning more than $1 million per week. He set up a charitable foundation and even became president of the Hong Kong Rotary Club. Fortunately, Benter was kind enough to contribute to the new center at CUHK. "Yau," he told me, "I've made money through math, and I want to give some money back to math too."

I was able to raise an even bigger chunk of change from Robert Kuok, a prosperous Hong Kong resident who controls the Shangri-La hotel chain. My childhood friend Ping Wah Chow—who lived near me in Shatin, Hong Kong, and later became an accountant to the Kuok family—introduced me to his boss. Kuok was extremely generous, becoming the largest single donor to the new math institute. He in turn introduced me to his friend Li Ka-shing, the richest man in Asia, who was born in Chaozou, China, the town that borders my birthplace, Shantou. Li also made a substantial contribution toward this endeavor. I managed to raise additional funds from Thomas Chen, the uncle of my friend Ronnie Chan, and from William Mong, who ran a Hong Kong electronics company. The Lee Foundation was kind enough to sponsor an endowed professorship, as did several other philanthropies. I'm used to dealing with big numbers as a mathematician, but being around all these millionaires and billionaires, in quick succession, almost made my head spin.

This fundraising took place over the course of several years and was essential for supporting visiting faculty and securing operating funds for the new center. Some people have said that I'm too blunt when it comes to asking for money, but I simply don't have the time, nor is it my style, to wine and dine potential donors. I spare people the niceties and just come right out and ask them. In this case, the direct approach paid off. The Institute of Mathematical Sciences (IMS) was established in 1993, and my friend S. Y. Cheng agreed to run things on a day-to-day basis as the associate director. (I've been the center's director from the beginning.)

Despite his outstanding credentials in mathematics, Cheng did not find the job easy from an administrative standpoint. He was beset with continual headaches from the CUHK dean until he finally got fed up and left. I eventually found a good replacement to keep the center running smoothly. Zhou-Ping Xin, who'd been a professor at the Courant Institute, became the IMS associate director in 1998 and has been there ever since. Many people don't realize that the biggest challenge in starting a new center is not only constructing and financing a building, but also getting good people to run the place.

IMS was the first center I ever formed. It has an active, degree-granting program for graduate students (with more than forty PhDs awarded so far), and it hosts postdoctoral fellows and visiting professors as well—many of whom come from the mainland. The center manages three international research journals, including *The Asian Journal of Mathematics*, which has published many important papers. Although we went through some rough patches in the early going, the center has, by most accounts, been a welcome addition to the Asian math community.

That all started in 1992 while I was wrapping up my appointment at Tsing Hua in Taiwan and my sons were finishing their school year in Hsinchu. I then took one more tour of China with the boys, which included a thrilling boat ride through the Three Gorges section of the Yangtze River, two years before construction started on the Three Gorges dam, which is part of the world's largest hydroelectric power project. Afterwards, we flew back to Boston.

I guess Michael eventually found some kind of answer to his question of why he had to learn Chinese. Years later, to my surprise, he took a course in Chinese literature (including ancient poetry) when he was an undergraduate at Harvard, following that up with a Chinese language program in Shanghai.

Meanwhile, I resumed my responsibilities as a Harvard professor. (I suppose all good things, even sabbaticals, must come to an end.) But even though I was nearly seven thousand miles from Beijing, my thoughts were not far from China and the additional projects I might still pursue there.

I called my friend Yang Lo, who was then president of the Chinese Mathematical Society, as well as a member of the Chinese Academy of Sciences, suggesting that China host the 1998 International Congress of

Mathematicians. Yang Lo liked the idea, later telling me it was well received among the leaders of China's science and mathematics community. Things proceeded quickly from there, although I knew that I would ultimately need Chern's approval too. Chern was initially skeptical, but S. Y. Cheng soon persuaded him that it was a good idea. The International Mathematical Union runs the congresses, and I contacted two former IMU presidents, Lennart Carleson and Jürgen Moser, who both got behind the plan.

Guoding Hu, the deputy director of the Nankai Institute of Mathematics (later renamed the Chern Institute of Mathematics), then used his long-standing ties to the Communist Party to arrange for Chern and me to meet with Chinese president Jiang Zemin to discuss plans for the ICM and scientific research in China more broadly. I flew to China in late April 1993 for the meeting with President Jiang, arriving several days early so I could spend time with Chern in Nankai and also attend a conference at Zhejiang University in Hangzhou—an event being held to honor the one hundredth anniversary of the birth of Jiangong Chen, a mathematician who specialized in Fourier analysis. Silei Wang, one of Chen's former students, who had also served as my assistant at IAS in the early 1980s, had organized the conference. I assured Wang that I would speak there.

The promised speech, however, did not happen because I was waylaid—and essentially held hostage—in Nankai, where I stayed in a guesthouse attached to Chern's home. When I asked people there to book a flight for me to Hangzhou, they took my passport, which would have been perfectly normal were it not for the fact that they did not return it to me. Without the passport in my possession, I could not fly to Hangzhou to attend the memorial conference for Chen. Chern asked me to stay in Nankai and speak with the other members of the institute, as they were in the midst of a special year in geometric analysis. But Chern would not let me move freely about the campus; instead, he offered to bring people to me, basically picking and choosing the mathematicians I was allowed to talk to. So I was, for all intents and purposes, a prisoner for those few days. Although I did not like being treated that way, I decided not to raise a big fuss because Chern had always been a father figure to me. Besides that, I did not want to jeopardize the upcoming meeting with President Jiang.

You might wonder why such extreme measures were taken to keep me from going to Hangzhou. Some people in Nankai, evidently, did not want me to pay tribute to Jiangong Chen. In fact, friends of mine at Zhejiang told me that Chern was behind the effort to keep me away. They claimed that even though Chen had died more than twenty years earlier, in 1971, Chern had not stopped competing with him. I don't know if this explanation was correct, but it sometimes seemed to me as if all the academic leaders in China were fighting with each other. And these feuds continued, in some cases, long after the original contestants had died.

Fortunately, Chern and I were able to ride together, peaceably, to Beijing, when it came time to meet with China's leader. We didn't discuss my involuntary confinement in Nankai during the drive, as Chern acted as if nothing out of the ordinary had happened. And if anything unusual had taken place, the inference was that he had nothing to do with it.

I was eager to see Jiang Zemin, which was obviously a great honor, even though I was not the sort of person who normally went out of his way to meet with political leaders. Yet as our van headed toward Zhongnanhai, the Communist Party's central headquarters, I thought back to an opportunity which had arisen several years earlier that I had regrettably passed up. During a visit to Beijing in 1986, Yang Lo told me that if I waited a week, I could meet with Deng Xiaoping, the "paramount leader" of China. Unfortunately, I did not have a week to spare, owing to pressing responsibilities back in the States, and that chance slipped by. Looking back, I realized that I should have found a way to spend time with this great leader, who had done so much to spearhead China's economic transformation in the latter part of the twentieth century.

But that was in the past. We presently had a two-hour drive from Nankai to Beijing during which I tried to collect my thoughts for the conversation with China's current president. Chern seemed nervous about our meeting with President Jiang though he was mainly preoccupied with funding for his mathematics institute in Nankai. He appeared to be much less concerned about the possibility of hosting a future ICM. His apparent indifference to that prospect might have owed to the fact that he was almost eighty-two years old then and was not sure he'd be around if and when an ICM was ever held in China.

But I tried to make a compelling case to President Jiang for the congress, noting that the country's math and science institutions had suffered badly during the Cultural Revolution and were in need of a major upgrade. An event like this, which would bring the world's top mathematicians to Beijing, could draw attention to mathematics as a field worthy of additional support. Mathematics professors were especially in need of support, I pointed out, given how low their salaries were. Jiang was well aware of that situation, mentioning that his own salary as president of China was just 800 yuan (less than $100) per month. Once I heard that, I did not press the matter further.

At the time of our discussion, China was making a bid to host the 2000 Olympic Games—which would cost billions of dollars. Why not spend a tiny fraction of that sum, I suggested to Jiang, just a few million dollars, to host the world's largest mathematical conference, which would attract the world's best mathematical minds to China? Fortunately, the president, who had a background in electrical engineering, embraced the idea. We had only a half hour scheduled with him, but he was in a talkative mood. We spoke for an hour and a half, and he had a lot to say about the importance of science and his desire to improve research in China.

My original thought had been to try to host the 1998 ICM in Beijing, but the International Mathematical Union decided to hold that session in Berlin, so we ended up applying for 2002 instead. (China, incidentally, failed in its bid for the 2000 Olympics but hosted the 2008 games in Beijing at a cost of more than $40 billion, according to the *Wall Street Journal*.) But even with President Jiang's approval, organizing the congress I had initiated soon became complicated—so complicated that by the time the conference took place, some nine years later, I'd been written out of the equation altogether.

Before getting into all that, I'll mention that the Chinese Academy of Sciences created a new membership category, Foreign Associate, in 1994. Chern and I were the only mathematicians picked for this designation the first time around. I wasn't able to attend the induction ceremony, but I met with the academy's vice president, Yongxiang Lu, a year later, in May 1995, when I gave a lecture at the sixtieth anniversary of the Chinese Mathematical Society in Beijing. I talked about the modern way of doing research, what China was doing wrong, and how it could be doing much better, following the example of the best American and European institutions. I warned, however, that it would not be easy because the country's research establishment had been dealt a serious blow during the Cultural Revolution, which left it lagging far behind its Western counterparts.

Afterwards, Lu told me that he had recorded my talk and was going to share it with the leaders of China. "I want you to help the academy," he said. He specifically wanted me to lead the formation of a new mathematical institute inside the academy, following the new approach to research I had outlined in my talk. "The old ways are not working very well," Lu admitted. "We have to overhaul the entire system, and we need your help in doing that." Having Lu on my team was very helpful, as he soon became quite influential in China, eventually rising within the ranks of the Communist Party to the level of vice chair of the People's Congress.

The next morning, I had a chance meeting that further helped the cause. At breakfast at the Peking Hotel I ran into Ronnie Chan, a real estate mogul I'd been friendly with since the 1970s. Chan seemed unusually animated. "It looks like the academic world is going to change for the better!" he said. I asked him what he meant, and he told me that the news story about my talk had appeared on page one of the *People's Daily*, whereas a meeting of Li Ka-shing and other top business leaders with President Jiang only made page two. "That means the government is starting to take your side and provide more support for academic research," Chan said.

I appreciated his enthusiasm, but appreciated even more the fact that Chan personally wanted to get involved in the mathematics research initiative. This was particularly good news because Lu had not yet lined up any funding for the new math center he hoped to create. In 1996, after several rounds of discussions, I reached an agreement with Ronnie and his brother Gerald Chan, who were both successful Hong Kong–based property developers. Ronnie was eager to fund the building. Gerald, who was more interested in research than his brother, said that the building alone was not enough; they should fund the center's first five years of operating expenses as well. Who was I to turn down such a generous offer? Lu then suggested that we call the new facility the Morningside Center of Mathematics, named after the Chan family investment company, the Morningside Group, and its charitable arm, the Morningside Foundation. Ronnie was especially excited by one aspect of my proposal, which is something we had discussed on several occasions: The center would regularly award—every three years, as it turned out—the Morningside Medal of Mathematics, which would serve as the Chinese equivalent of the Fields Medal.

A groundbreaking ceremony was held on June 10, 1996, and the academy's president, Lu, was there, lending his gravitas to the affair. I spoke at the ceremony, claiming that this would be the first "open" mathematics center in China, meaning that every qualified person could apply to study there. People would come to do research for a year or less and then return to their home institutions. That way other places in China would not have to worry about losing their best mathematicians.

We invited Chern to the groundbreaking ceremony, but he chose not to attend. I had discussed the center with him on several occasions, and he always maintained that he didn't have any problem with it. Nevertheless, I heard that he was angered by it. And that, unfortunately, was the way Chern often operated, not telling you directly how he felt, while complaining bitterly to others about what he saw as your less than honorable conduct.

I gave the dedication speech at the ceremony, noting that

> we mathematicians are not looking for fortune, nor long-lasting dynasties, since all these will eventually turn to ashes. We are seeking theories and equations that can carry us forward on the path towards eternal truth. These ideas are more valuable than gold and more radiant than poems, both of which pale in comparison to the naked truth. Mathematical strength can make countries rich and powerful, since that knowledge forms the basis of all applied sciences. A mastery of mathematics can also help keep countries in peace, owing to the essential role the discipline plays in the planning for, and maintenance of, a modern society.

Kung-Ching Chang, an influential mathematician from Peking University, also said a few words at the ceremony. The basic message he conveyed was that he would do whatever it took to move the center to Peking University. That curt utterance, which had a chilling effect on the proceedings, grew out of complaints he and his Peking colleagues had raised about the money allocated for the Morningside Center. It was their standard position that Peking University, as an institution of unparalleled excellence, should be entitled to half that money—and, indeed, half of any money given out on behalf of mathematics. There was no particular reason for this expectation; that was just their usual stance. But the Chan brothers were not interested in funding a math center at Peking University, nor had that been my intention either. We wanted the new center to be installed at its intended home, the Chinese Academy of Sciences, which was where China's strength in mathematics was already concentrated and where the academic atmosphere was more congenial, at least to my tastes.

Our collective refusal merely served to enrage the Peking University crowd further, who joined forces with Chern and others at Nankai University and called for Lu's resignation because he was unwilling to share resources with them. The university wielded a lot of clout in China because many of its graduates were ministers in the government, but it was not strong enough to defeat the Academy of Sciences. Peking University partisans had additional complaints: By naming our center after a private company, they claimed, we had betrayed China. That charge struck me as hypocritical since they gladly would have taken half the money from this private concern had we given in to their demands.

A friend of mine, Jian Song, had been planning to attend the groundbreaking event. Song was an academician at the Chinese Academy of Sciences, as well as a prominent government official at the vice premier level. But at the last minute, he was persuaded not to come by lobbyists from Peking University who maintained that his presence would constitute an endorsement of the center and would therefore be held against him.

This is yet another example of a lesson I've learned, the hard way, on all too many occasions: Dealings in the United States can be complicated and frustrating, but in China they can be far worse. Fortunately, in this particular skirmish, we had a formidable ally, President Jiang Zemin. He dismissed the claims made by Peking University, and with his backing, the Morningside Center went ahead as planned—albeit with the inevitable bumps in the road.

At the time of the groundbreaking ceremony, of course, we still didn't have a building, nor had we finalized the designs for the center. The Chan brothers and I had many deliberations of an architectural nature. I made the astute suggestion that "this building cannot look lousy." Ronnie Chan must have taken that comment to heart, as he hired a first-rate architect with whom he'd had good success in the past. The center eventually won an award as one of the most elegant buildings of its size in Beijing. Nevertheless, significant issues arose before that happened, including a dispute over the bathrooms. The builder was trying to save money by putting in the old-fashioned toilets that do not allow one to sit down; instead, one must squat over a hole in the floor. In the end, most of the building's toilets were of the modern variety, but the first-floor bathrooms featured squat toilets. Ronnie and Gerald's mother refused to use those bathrooms, and I personally found it embarrassing to have throwbacks like that in what we hoped would become a modern, world-class research facility.

By 1998, construction was completed and we were ready to begin the center's operations. I was the first director and have continued in that capacity to this day. Ever since the groundbreaking, however, I had been working on hosting an international meeting at Morningside, open to all mathematicians of Chinese ancestry. The Nobel Prize–winning physicists C. N. Yang and T. D. Lee had done something like this a decade earlier with Chinese physicists—an event attended by then President Deng Xiaoping—and I thought the same idea made sense for mathematicians. Yang Lo liked the idea too, so I wrote to K. C. Chang, who was then president of the Chinese Mathematical Society, thinking it would be helpful to have the society's support on this.

Chang told me that the society would get behind the conference, but he insisted on having complete control of it and all future conferences of this sort, should this become a regular event (as did, in fact, occur). Chang further insisted on inviting all of the speakers. I, on the other hand, had no opportunity to "insist" on anything because I was not asked to give my opinion. I did have a strong opinion, nevertheless, the key point for me being that speakers should be selected on the basis of academic merit. The Peking University group wanted speakers throughout China to be recognized, regardless of the quality of their work. The main objective of such a conference, to them, seemed to be to make everybody happy. That was not my idea of how to promote mathematics in China. My whole life has been about academic quality, and I told them that if they wanted to throw a big party, they could do it themselves.

Predictably, that remark angered them. The Chinese Mathematical Society, which Chang headed, issued two statements asserting that an international conference for mathematicians of Chinese descent should not be held unless such a conference were run by the society itself. Chang even lobbied the IMU to intervene. An American mathematician who used to be active in the IMU asked me why I wanted to run this conference. Maybe it would be more appropriate for the IMU or Chinese Mathematical Society to run it, he suggested. I told him that I had run many conferences at Harvard and elsewhere—some of which he had attended—without asking the American Mathematical Society to get involved. "You guys are always talking about freedom of speech in China," I said. "Do you now want the Chinese Mathematical Society to step in and squelch academic freedom?"

Chang then asked Yang Lo to tell me that the Chinese government did not want me to run such a conference. I told Yang Lo that I would cancel the event, or my involvement with it, if I received a formal letter from the Chinese government making the request. I was confident that no such letter would arrive, and it never did. Once Chang knew that I was going ahead with a conference, he wrote a letter to the Chinese Mathematical Society advising all of its members not to attend. His letter essentially told the whole country not to attend.

The first International Congress of Chinese Mathematicians (ICCM) ran from December 12 to 16, 1998, and subsequent congresses have been held every three years. The first-ever opening ceremony took place in the Great Hall of the People in downtown Beijing, about an hour from the Morningside Center. On December 12, when the conference kicked off, a dozen buses picked up more than four hundred congress participants from Morningside, carrying them in a "great convoy" to the Great Hall for the awards festivities.

The Morningside Medals are traditionally given out on the first day of the ICCM. Two gold medals, each with a cash prize of about $25,000, and four silver medals, each with a cash prize of about $10,000, are bestowed to mathematicians of Chinese descent who are forty-five years of age or younger. The Fields Medal has an age limit of forty, but I wanted to extend that because people can do outstanding work in their forties. (Andrew Wiles, for example, completed his revised proof of Fermat's Last Theorem in 1995, and that momentous achievement took place when he was forty-two years old.) Chang-Shou Lin, of the National Chung-Cheng University in Taiwan, and Shou-Wu Zhang, who was then at Columbia University, were the first gold medal winners.

The selection committee was made up of non-Chinese mathematicians, except for me, in the hopes of keeping internal politics out of the deliberations, and that policy worked well. There was no dispute over the winners, although I heard that some mathematicians from Peking University were disappointed that Gang Tian did not get the award. But I had decided, as a matter of principle, not to give any prizes to my students, past and present, the first time around, so as not to show signs of favoritism.

Nor did Tian get an award at the second ICCM, which was held in 2001. The gold medals that year went to Jun Li, my former student, and to Horng-Tzer Yau (now my colleague at Harvard)—both of whom I and other panel members felt had done superior work.

Chern received a Morningside Lifetime Achievement Medal at the second congress in 2001. Chern's daughter accepted the award for him, as his doctor had advised him not to travel for health reasons. Although Chern had initially protested the ICCM, complaining about the idea in several letters to China's president, he later changed his mind and decided he wanted to attend the next congress. He also began to support the ICCM, going so far as to make a sizable cash contribution. Such reversals were not uncommon for Chern, though he often did not explain why he changed his mind on one point or another.

On balance, I consider the formation of the ICCM a resounding success, and I believe that most conference participants would agree. In my opening remarks in 1998, I called the congress a "historic event—the first occasion where the majority of Chinese mathematicians from all over the world gathered together to present their research." Several distinguished non-Chinese guests attended as well, including Ronald Graham, the former president of the American Mathematical Society; Jean-Pierre Bourguignon, president of the European Mathematical Society and the director of IHES; Jürgen Jost, director of the Max Planck Institute for Mathematics; and Martin Taylor, president of the London Mathematical Society.

The Morningside Center has been another success. I don't think I'm alone in regarding it as one of the best mathematical institutes in China. I went on to found several more mathematics centers in China—at Tsinghua University in Beijing, at Zhejiang University in Hangzhou, and at Sanya in Hainan. There was also the aforementioned IMS in Hong Kong, along with the new centers I started in Taiwan, first at National Tsing Hua

University and later at National Taiwan University. But in the early battles to establish the Morningside Center, as well as to form the ICCM, I made many enemies. In fact, there have been big fights over every center I've been involved with—in China, Hong Kong, and Taiwan.

Around 2010, Michael Atiyah called me, seeking my opinion on whether he should take the position of director of the Tsinghua center in Beijing. Apparently, C. N. Yang had asked the renowned mathematician to become the center's director—a rather astonishing move, given that I was the center's founding director and its acting director, and I had heard no hints regarding any changes in leadership. Furthermore, it's not clear that Yang had the authorization to do this, as Tsinghua's president told me he knew nothing about the so-called offer that Yang (who now serves as honorary director of the university's Institute for Advanced Study) had made to Atiyah. I would have approved of having a mathematician of Atiyah's stature step in and take an active role in building up Chinese mathematics, though Atiyah informed Yang that he would be able to spend only about one week each year in China. Atiyah told me that Yang thought that was fine, but I don't think that would have been a good idea, knowing from experience that one cannot effectively run a center by "remote control."

As best I can tell, this idea was dropped after I spoke with Atiyah. What I know for sure is that I've maintained my post at Tsinghua University's Mathematical Sciences Center ever since its founding in 2009, and I've enjoyed the administration's support the whole time. In 2015, moreover, the Chinese Ministry of Education designated the center a national research institute, changing its official name to the Yau Mathematical Sciences Center—a move that left my standing seemingly secure. So apparently this incident with Yang was just one of those random blips on the screen, which came and went without leaving a lasting mark.

Meanwhile, fights over the ICCM persisted for years, long after the congress had established itself as the most widely attended Chinese math gathering. One contributing factor, it seemed to me, was that a group of people at Peking University didn't like being overshadowed. They continued to call for a boycott of the ICCM, or to shut it down completely, long after it had become the biggest mathematics conference in the country. They abandoned the boycott only after I showed them a letter from the minister of education, Zhili Chen, praising the event. For at that point, they had run headlong into a harsh reality of Chinese politics: It is all but impossible to contest a statement made by a Chinese leader, which is normally issued after consultation with other government authorities. In this case, the system worked, as Minister Chen's recommendation in support of the ICCM was (in my admittedly biased opinion) quite sound.

With the ICCM now on firm ground, my rivals needed another issue to fight over, and they found it in the proposal, first advanced by me, for China to host the aforementioned International Congress of Mathematicians (ICM—note the presence of just one C in this acronym). This proposal had moved ahead, and the IMU had approved a plan for the congress to be held in Beijing in 2002. My original hope had been that this conference would provide a big boost to the Chinese mathematical community, but it instead set off a struggle for power and influence that was the opposite of what I had intended. I was soon relegated to the sidelines before the whole thing came to pass.

The IMU decided that eight mathematicians, selected by the Chinese Mathematical Society, would have an opportunity to speak at the congress. As usual, I insisted that speakers be picked on the basis of consequential new work they had carried out, but my rivals made sure that I had no influence over the proceedings. Meanwhile, everyone was competing to give a talk at the conference, as well as to be on the committee that organized it. What happened in the end was not surprising: Lecture spots were given out on the basis of political connections rather than academic achievement, and I had no input in the process.

The stakes were quite high. Giving a talk at the ICM meant instant recognition, money, and prestige. It meant that your academic department would recognize you as one of the major people working in the field, which would almost automatically lead to a promotion and perhaps a special award highlighting the fact that you were a special person—a force to be reckoned with.

It was not until these eight slots were given away that the Chinese Math Society members who made these decisions asked me to have a hand in organizing the congress. They were correct in assuming that had I participated earlier, I would have interfered with their choices. By this time, I was disgusted with how things had gone and disinclined to have anything further to do with the event.

Even though they had wanted me out of the way, the organizers realized it would not look right if I skipped the proceedings altogether. The Chinese government also wanted me to be there, and some officials asked Chern to persuade me to attend. The situation was somewhat comical in that I'd heard from several sources that Chern had already told President Jiang Zemin that I should *not* go to the congress. Nevertheless, I acceded to the request to meet with Chern at Nankai University, where we had lunch and spent several hours together. During that time, Chern did not say a single word about my going to the congress. Afterwards, I imagine that he told others he had done his best but was unable to persuade me.

A year or so before the event, Zhi-Ming Ma—the chair of the local organizing committee for ICM 2002 and president of the Chinese Mathematical Society—wrote to me, saying that he would be in the United States and wanted to talk about the congress. I didn't hear from him further until he contacted me at the last minute, mentioning that he'd be in Boston the next day and would like to meet. My hunch is that he hoped I would be busy, but I invited him for dinner. We dined together, but he never mentioned the congress or my possible role in it.

By 2002, the local organizing committee felt pressure from the IMU to confirm my presence. I was told that IMU president Jacob Palis wanted me to deliver a plenary lecture, but all those slots had already been handed out. Under duress, the organizing committee asked me to give a special talk after dinner.

Again I wrote to them, expressing my dismay over the fact that the choice of speakers was so closely tied to politics, which I felt should have no place at the congress at all. But that's not what they wanted to hear. The letter I got in return conveyed the message—in tone, if not in words—that I should not attend. Although the Chinese government wanted me there, as did the president of the Chinese Academy of Sciences, many others, who wielded some influence, were not so welcoming.

In the end, the event that I had conceived had become so tainted that I decided to skip it, although I did not discourage others from attending. In the meantime, I focused my energies on an international string theory conference I had organized, which took place in Beijing from August 17 to 19, 2002, just before the ICM. I got some prominent people to participate in this forum—including Stephen Hawking, Edward Witten, David Gross, and Andy Strominger—and that in itself made for some big news. More than two thousand people came to hear Hawking's public lecture. I introduced him, along with Witten, Gross, and Strominger, to President Jiang, who was eager to send Chinese scholars to study string theory in the United States.

The three-day conference was gratifying to me, as it brought together math and physics, and East and West—two important causes that I have devoted considerable energy toward. I was also happy to see more than two hundred researchers from all over the world convene at such a high-profile affair, held in my native country with some significant media attention.

The International Congress of Mathematicians started on August 20, a day after the string theory gathering wrapped up, and as noted, I missed that altogether. The congress had simply rubbed up against too much ugly politics for my tastes. Politics crops up in every country that hosts the ICM, but China, it seems to me, is particularly bad in this regard.

Another thing that is unfortunate in China is the number of fabricated stories that are posted anonymously and disseminated on the web. (This problem is not limited to China, of course.) One widely circulated story held that I asked the Hong Kong Mathematical Society to write to the IMU to stop the congress from being held in Beijing for human rights reasons because I wanted it to be held in Hong Kong instead. This claim was total bunk. S. Y. Cheng, who was then president of the Hong Kong Mathematical Society, has vouched for the fact that I never made such a request. The Hong Kong society had already sent a letter lending its enthusiastic support for the notion of having the congress in Beijing. So the charge about my desire to change venues was simply part of the smear campaigns that people often have to endure if they are any sort of public figure in China.

Although I saw Chern occasionally after the 2002 congress, we never managed to settle our differences. That said, we still agreed on much. We both loved China and we both wanted to boost the level of mathematics carried out on the mainland, though we had different ideas about the best ways of realizing this objective. Chern was in a hurry (perhaps because of his age) and therefore focused more on short-term goals, whereas I favored a more methodical, long-range strategy aimed at establishing a high-quality research environment. This takes time to build up; I'm not aware of any shortcut to excellence.

Sadly, the one thing Chern and I did not have was time, which might have given us the opportunity to reach a common ground, because in the end we wanted the same thing. But fate intervened. In early December 2004, I received a call from Yang Lo telling me that Chern had died at the age of ninety-three.

I felt deep regrets that our relationship had gotten so badly off track. I wish we'd been able to make peace with each other. Now that he was gone, I tried to remember all the things he had accomplished and how grateful I was for how much he had helped me early in my career, including smoothing the way for me to go to Berkeley in the first place. Back then, Chern loomed as a larger-than-life figure. Visiting him in his office to seek assistance was for me, a young graduate student, kind of like asking for a favor from Don Vito Corleone, as portrayed by Marlon Brando in *The Godfather*.

I am still filled with admiration for Chern's prodigious achievements in mathematics. He truly was one of the principal founders of modern differential geometry. I gave a tribute to him in the opening talk of the 2004 ICCM, which was held in Hong Kong two weeks after his death and was dedicated to his memory. I even recited a poem I had written about him. Unfortunately, the room my talk was held in could accommodate only 250 people; there was not nearly enough space for the throngs of people who wanted to be there.

According to those who were close to Chern near the end, just before he left this world he said that he was "going to see the Greek geometers." I have no doubts that he would have more than held his own in that crowd. Chern's contributions will endure, just like those of Pythagoras and other legendary figures in the history of mathematics. In fact, the International Astronomical Union named an asteroid (discovered at China's Zinglong Observatory) after Chern as a way of honoring all he did to advance the field of mathematics.

Chern never lost his passion for the subject, continuing to pursue his discipline as far as his energy and faculties could take him—and far beyond any normal retirement age. Part of the impetus might have been his competitive spirit, which was still going strong. But the main driver, I assume, stems from the fact that he loved mathematics and simply could not let it go.

In 2003, Chern was still working hard, including on a proof of an important problem dating back more than half a century concerning the six-dimensional sphere or "six-sphere." He asked me to comment on his paper. I could see that his proof did not hit the mark, though I told him I admired the fact that someone in his nineties could take on such a challenging project. He seemed to be pleased by that remark. One mathematician called this effort "Chern's last theorem," but Chern was in fact hard at work on something else as well, almost until his very last breath. Colleagues at his institute in Nankai said that the light in his office was almost always on. Chern thought he had proved the Poincaré conjecture —when he was almost as old as the problem itself—with a novel approach that rested upon fairly simple calculations. Those calculations did not look convincing to me, and by now, more than fifteen years later, it's probably safe to assume that Chern's late-career proof did not add up, as no one—as far as I'm aware—has ever made use of it or provided any reason to believe that it could be correct.

After Chern's death, his son-in-law Paul Chu, a physicist who served as president of the Hong Kong University of Science and Technology, was upset that Chern's "proofs" regarding the six-sphere and Poincaré conjecture had never appeared in print. Chu complained to me and my friend S. Y. Cheng, who was then the dean of science at HKUST, about the fact that we had not found a publisher for this work. Both Cheng and I were reluctant to intervene in this case, as we felt that Chern's latest offerings—being well below his usual standards—might tarnish his legacy. I felt that Paul Chu, as a member of the family, should pursue this himself since he had rights to the work. I still think that Cheng and I did the right thing, in terms of protecting Chern's reputation, though our suggestions did not sit well with his family at that time.

And although in my opinion Chern's final two theorems fell short, I was still impressed that he had taken on such immensely daunting projects at such a late stage in his life. He kept at his work gamely, until he could do no more.

All told, Chern put together an amazing career in mathematics—leaving behind a rich body of work for others to carry further, as well as an asteroid bearing his name, which continues to travel around the sun in an orbit, or curve, that shall forever remain elliptical.

CHAPTER ELEVEN

Beyond Poincaré

"ROSE IS A ROSE IS A ROSE," Gertrude Stein famously stated in a 1913 poem. But can the same be said about a sphere? If you take, for instance, a slightly deflated soccer ball and push it at one end, pull it at another end, step on it, jump on it, twist it, pummel it, and do everything else imaginable save for poking a hole in it or ripping it apart, will this ball remain a sphere insofar as the dictates of topology are concerned?

The brilliant French polymath Henri Poincaré—who made important contributions to wide-ranging areas of mathematics, as well as to celestial mechanics, the special theory of relativity, and other realms of physics—posed a similar question in 1904. Couched in more technical language than Gertrude Stein employed, Poincaré's question took the form of a full-fledged mathematical conjecture, surely one of the most talked-about conjectures of all time. It stood for nearly a century, withstanding scores of failed attempts at solving it, until the first credible proof emerged, seemingly out of the blue, in a series of web postings by the Russian mathematician Grisha Perelman in late 2002 through mid-2003.

What is this conjecture that has attracted so much attention over the past one hundred years and remains, to this day, a politically charged subject? As mentioned in Chapter 5, Poincaré asserted that a compact space—within which no two points can be infinitely far apart—is identi-

A sphere, consisting of a two-dimensional surface, is regarded as "simply connected" because any loop placed on the sphere's surface can be shrunk down to a point; no obstacle stands in the way.

cal to a sphere, from a topological standpoint, if any and every closed curve you can place in that space can be shrunk to a point. No obstacle, in other words, would prevent any such loop, put anywhere in that space, from contracting down to a single point. I've long marveled at the fact that this conjecture is so short. It can be summed up in a single sentence, yet that sentence kept people busy for more than a century, which is one of the reasons I find Poincaré's formulation so beautiful. (Please bear in mind that the above description only applies to the three-dimensional version of the conjecture. In higher, n-dimensional statements of the problem, the condition to be satisfied involves shrinking spheres [of dimension less than n] to a point, rather than shrinking circles or loops.)

It's probably easiest to picture what Poincaré had in mind by looking at a so-called 2-sphere, such as the two-dimensional surface of a world globe (though not the *interior* of that globe). You could, for example, stretch an exceedingly tight rubber band around the equator and gradually slide the band toward the north or south poles, where it would encounter no impediment—nothing to stop it from contracting to a point. On the other hand, a band wrapped around the middle of a donut—a space that, by definition, contains a hole—could not contract to a point unless either the band or the donut was torn. A band wrapped around the outer or inner circumference of the donut could be shrunk down only if the donut were squished and the hole thereby eliminated (so that the object in question could no longer rightly be called a donut).

And, again, remember that we are talking about just the *surface* or shell of that donut, not its (sometimes tasty) interior. The key difference between these two shapes is the presence or absence of a hole running through the surface, which a donut notably has, whereas a sphere does not. This means that a sphere cannot be transformed into a donut without

Three possible loops can be placed on the surface of a torus, or donut—only one of which, the small loop on the right, can be shrunk down to a point. A torus, therefore, is not "simply connected."

A loop stretched around the outer rim of a torus cannot be shrunk to a point because the hole in the middle keeps it from shrinking.

This loop cannot be shrunk down to a point either without cutting the torus, thereby changing its topology.

tearing the fabric of its internal space, nor can a donut be transformed into a sphere without repairing that tear.

Since two-dimensional surfaces were well understood in the nineteenth century, Poincaré's conjecture specifically concerned 3-spheres, such as the surface of a four-dimensional ball—spaces that are not easy for most people to visualize. Just as a 2-sphere consists of the set of points that lie exactly the same distance—call it r—from the origin in a three-dimensional space (satisfying the equation $x^2 + y^2 + z^2 = r^2$), the 3-sphere consists of the set of points lying equidistant from the origin in a four-dimensional space (satisfying the equation $x^2 + y^2 + z^2 + w^2 = r^2$). Understanding this theorem would lead to a much deeper understanding of three-dimensional spaces in general. Poincaré, however, was prescient enough to realize the challenges involved in such an undertaking. "Mais cette question nous entraînerait trop loin," he wrote, suggesting that the process of answering this question could lead us on a very long journey indeed.

The two-dimensional version of the problem was solved well before Poincaré presented his conjecture. Higher-dimensional versions were solved in various stages. In 1962, Stephen Smale proved the conjecture for all dimensions greater than four. Michael Freedman presented his solution of the four-dimensional case in a 1982 paper published in the *Journal of Differential Geometry* (see Chapter 7). But the three-dimensional case turned out to be much more obstinate, just as Poincaré had predicted. It's more difficult in part because of the fact that tricks employed in higher-dimensional proofs can't be used in three dimensions because there's simply not enough room to move around.

The three-dimensional problem, accordingly, served as the wrecking ground for countless failed proofs—the mathematical equivalent of the Bermuda Triangle in the Atlantic Ocean where many airplanes and ships have met their ends. The mathematician John Stallings—who proved the conjecture in 1960 for dimensions greater than six—recounted his own failed attempt at glory in taking on the three-dimensional problem in a 1966 paper titled "How Not to Prove the Poincaré Conjecture."

I've long been interested in the conjecture, and my thoughts have turned to it from time to time, including, as I said, during the cross-country drive I took with Yu-Yun and my in-laws just before our wedding in 1976. I never committed myself fully to this enterprise mainly because

I never had that great moment of inspiration that could crack the case wide open. Poincaré once described thought as "a flash in the middle of a long night. But this flash means everything." When it comes to this problem, I sadly never experienced such a "flash."

Instead, my role has been more of a supporting nature. Richard Hamilton has claimed that "no one is more responsible than Yau for creating the program on Ricci flow, which Perelman used to win this prize [the Fields Medal]." That statement, generous as it may be, clearly goes too far. No one was more responsible than Hamilton himself for creating the overall approach, or foundation, from which Perelman's argument grew. My contribution was in helping Hamilton develop this program because I recognized from the start the great promise that this work held.

Ricci flow (see Chapter 7) is a technique developed almost exclusively by Hamilton. This approach, which is based on differential equations rather than on standard topological methods, provides a geometric analogue of heat flow. If, for instance, you take a big slab of metal and direct a blowtorch toward a small portion of it, that part, naturally, will get quite hot. But if you then leave the slab alone, heat will gradually diffuse from the hot spot, spreading throughout the metal until the object reaches thermal equilibrium, achieving a uniform temperature at every spot on the surface.

Ricci flow is similar in terms of being an averaging process. But instead of evenly distributing heat, it tends to smooth out bumps and irregularities on geometric spaces: Regions of high curvature are gradually transformed into regions of lower curvature until the space eventually achieves uniform curvature throughout—just like a sphere, which is a surface of constant (positive) curvature. Some bumps, however, are more intransigent and can't be ironed out so easily. Instead, creases and folds, which mathematicians call "singularities," may form, requiring more drastic measures (which will be discussed shortly).

Hamilton initiated this project in the early 1980s, and I met with him regularly to discuss pressing issues that had come up in the course of his research—offering suggestions, pointing to some relevant work I'd done earlier with Peter Li, and generally supporting and encouraging him to the extent that I could. I also sent students to learn from him, work with him, and, I hoped, help him on his decades-long endeavor. I may not have been the first to make such an observation, but I did tell Hamilton back in the 1980s that Ricci flow might well be the key to solving the Poincaré conjecture. Although the suggestion might have been obvious, Hamilton was excited to hear it said out loud. The principal challenge, I told him, would be to understand the number and shape of singularities that could arise during the process of Ricci flow.

The mathematics here, involving some elaborate differential equations, is extremely demanding, and the number of researchers who have mastered it is still quite small. Yet the basic strategy can be described in relatively straightforward terms: You take a somewhat roundish object, apply Ricci flow, and find out whether—in the course of evening out the curvature—it can evolve into a sphere. With three-dimensional surfaces, especially highly irregular ones, the process can run into snags, resulting in the formation of singularities. Most of them can be removed, and literally cut out, through "surgical" techniques—like those introduced by John Milnor—a viable remedy as long as the number of procedures to be performed is limited.

One type of singularity, a cigar-shaped protuberance, could not be eliminated through surgery. And during the process of Ricci flow, when curvature tends to be averaged out, this sort of outcropping, rather than getting smoothed over, could grow uncontrollably large. These so-called cigars, Hamilton declared, posed the biggest obstacle to proving the Poincaré conjecture because if they were to form, it might be impossible to achieve a uniform geometry—and thereby establish a space's equivalence to a sphere—through Ricci flow techniques.

If, on the other hand, Hamilton could show that the dreaded cigars would not appear, the problem could be bypassed altogether. In fact, he explicitly demonstrated around 1996 that he could prove the Poincaré conjecture, assuming that cigar singularities would not sprout up and standard surgical procedures would apply. I suggested to Hamilton, and he agreed, that a way of taming these singularities, and assuring that they would not emerge, might lie in a powerful inequality stemming from the heat equations Peter Li and I had developed in the 1970s and 1980s. Hamilton started to explore this idea, and more than a decade later he enlisted my help in adapting the so-called Li-Yau inequality to the more general and more demanding conditions needed to address the Poincaré problem.

The curvature of a bumpy, three-dimensional manifold can gradually become smoother and more uniform during Ricci flow (*top, left to right*). Mathematicians had concerns, however, as to whether this process might go awry. In particular, the manifold might stretch out during Ricci flow to form an elongated "singularity," as the width of the neck connecting the two lobes goes to zero (*bottom*). (Figures courtesy of Christina Sormani, City University of New York.)

At a Harvard mathematics faculty meeting in 1996, I persuaded my colleagues that Hamilton was making headway on the Poincaré and geometrization conjectures. That effort, I said, could be furthered by having him at Harvard—a move that could benefit both him and the university. So Hamilton spent a year at Harvard as a visiting professor, starting in the fall of 1997. We talked regularly during that year and continued to exchange ideas afterwards. I went to Hawaii a couple of times, where Hamilton normally spent his summers. When we weren't working on Ricci flow, he enjoyed the flow of the Pacific Ocean while surfing. I enjoyed the beach too, though in less dramatic (and adventurous) fashion.

I also got an Eilenberg Visiting Professorship at Columbia in 1999, which gave me more time to work with Hamilton. Later, I even got a job offer from Columbia, which I turned down after talking things over with my wife, but Hamilton and I remained in close contact. He was making steady progress, with some help from me, and it seemed like the end might be close at hand. Then, on November 12, 2002, I received an unexpected e-mail from Grisha Perelman, someone with whom I'd had only minimal contact. Hamilton received a similar (if not identical) note around the same time. "May I bring your attention to my paper," Perelman wrote, referring to an entry he had posted a day earlier on the mathematical archives, "The Entropy Formula for the Ricci Flow and Its Geometric Applications."

The paper, which offered, according to Perelman, "a brief sketch of the proof of [the] geometrization conjecture," caught me, and probably most of the mathematical community, by surprise. I had no idea that Perelman was working on that conjecture, which is broad enough to encompass the Poincaré conjecture as well. The previous work for which he was best known had been in an entirely different realm of geometry. In fact, he had submitted one of his more noteworthy papers to the *Journal of Differential Geometry*, which I was then editing. We had fruitful exchanges about this work, and Perelman closely followed requests from referees to provide more details of his proof.

Often described as a recluse, Perelman left Berkeley in 1995, returning to his home in St. Petersburg, Russia, where he kept such a low profile that most of us did not know what he was up to or whether he was doing mathematics at all.

His first of three papers on "the entropy formula" was just thirty-nine pages long and appeared online on November 11, 2002. It was followed by a twenty-two-page paper, "Ricci Flow with Surgery on Three Manifolds," posted on March 10, 2003, and a short seven-page addendum ("Finite Extinction Time for the Solutions to the Ricci Flow on Certain Three-Manifolds") to the second paper, which came out on July 17, 2003.

The three papers were quite short by conventional standards for a proof of this breadth and complexity, and in them Perelman was able to demonstrate that the singularities Hamilton most dreaded would not form during Ricci flow. That was a dazzling piece of work—surely one of Perelman's greatest achievements—for "ruling out cigars," Hamilton said, was "the one part of the singularity classification that I could not do." By inventing new techniques for controlling these singularities, Perelman paved the way to a proof of Thurston's geometrization conjecture (mentioned in Chapter 7), of which the Poincaré conjecture was merely a special case. Moreover, the importance of these results—which are valid in *all* dimensions, not just in three—could, according to Hamilton, extend well beyond these conjectures.

Thurston postulated that three-dimensional space could be divided into eight basic shapes of uniform geometry, one of those being a sphere. Proving his geometrization conjecture would show how to carve out a sphere from three-dimensional space, as well as demonstrate exactly what goes into making such an object. In this way, a proof of the geometrization conjecture would contain a proof of the more narrowly defined problem concerning spheres alone that Poincaré posed, while also providing a thorough characterization of the geometry of three-dimensional space as a whole. (The conjecture had previously been proved for six of the eight shapes posited by Thurston—the spherical and hyperbolic cases being the two that, before Perelman, had eluded mathematicians.)

In his papers, Perelman promised to deliver an "eclectic proof" of the geometrization conjecture, rather than a fully fleshed out treatise with every step thoroughly delineated. He used a kind of shorthand in tracing out the argument, leaving out many technical details that he perhaps considered superfluous. Others, however, would not necessarily find the omitted particulars so obvious, even those with considerable expertise in this area. Although Perelman's proof was not easy to follow at every turn, it was clear to me that his papers did represent a tremendous advance in our understanding of three-dimensional spaces and the structure of singularities that can arise therein. There was no doubt that he had done an impressive bit of mathematics.

In April 2003, Perelman came to the United States to discuss his proof at MIT, Stony Brook, Princeton, and Columbia. At the time of his lecture tour, other mathematicians only had a month to digest his second paper, and his third paper did not come out until mid-summer. By then, Perelman was back in Russia and essentially out of contact with his peers. He told the British newspaper, the *Sunday Telegraph,* that all of his thoughts on the Poincaré conjecture and the broader geometrization conjecture were contained in his three online papers. "It's all there," he told reporter Nadejda Lobastova. "I have published all my calculations. This is what I can offer the public." And, as far as I know, he has said and written nothing more on this topic.

Perelman never published his papers in a print journal, which may well have asked him to elaborate on various points. I wrote to him several times, requesting that he publish this work in the *JDG*, which I'd been editing since 1980, but he did not reply.

It was therefore left to other mathematicians "to connect the dots" in Perelman's arguments—as the *New York Times* put it—to determine whether they were complete or contained any significant gaps, and to figure out exactly what had been established in them. I asked my former graduate student, Huai-Dong Cao at Lehigh University, and a former postdoc, Xi-Ping Zhu at China's Sun Yat-sen University, to comb through Perelman's papers and try to reconstruct the argument in full. I felt that Cao and Zhu might be up to the task, being far more qualified than most because of their extensive experience in Ricci flow methodology, which they started working on in the 1990s.

The Clay Mathematics Institute—a nonprofit foundation then based practically across the street from Harvard—also offered financial support to two teams of mathematicians to trace out the key steps in Perelman's proof: Bruce Kleiner and John Lott, as well as John Morgan and Gang Tian (my former student, with whom I was then at odds). The Clay Institute had an almost proprietary interest in Perelman's work because in 2000 it had listed the Poincaré conjecture as one of its seven "Millennium Prize Problems." The first person to submit an accepted proof of one of these problems, which had been published for at least two years, would receive $1 million from the institute.

Given that a proof of the Poincaré conjecture would constitute a huge milestone in our field, I was in favor of having as many mathematicians review Perelman's papers as possible. Being "old school," however, I've long held that the burden of proof should fall to the author rather than being left for others to verify. In fact, I grew up thinking it was one's duty as a mathematician to spell things out to other people, as well as to oneself, because until one writes a proof down completely, laying bare every single step, one is never sure that it is right. As you may recall, I had learned this the hard way in 1973 when I thought I had proved the Calabi conjecture wrong. But I found out three years and a good deal of embarrassment later that the conjecture was correct after all.

I also had a problem with the whole notion of the Millennium Prizes, because I didn't see that the Clay Institute had a legitimate claim to the Poincaré conjecture and the other well-known problems on its list, including the Riemann hypothesis. I felt that these problems were of historic value, belonging first to their authors and subsequently to all of mathematics. People should not be motivated to work on them simply because a foundation offers a big cash prize. Solving a problem of this magnitude should be its own reward, and I don't believe that any additional inducement is necessary. Nor do I believe that a foundation, however well heeled, has the right to appropriate some of the field's most pressing questions and attach its name to them. Perelman may have shared some of my misgivings about placing a monetary value on the Poincaré proof. In any case, he turned down the $1 million Clay prize when it was eventually offered to him, telling Russia's Interfax news agency that the award was unfair, given that his contribution to the conjecture's resolution was no greater than that of Hamilton.

I was not content to leave the vetting of a proof of this magnitude to the Clay Institute, which, in some sense, had a financial stake in this problem. So at my invitation, Zhu visited Harvard during the 2005–2006 academic year, where he lectured for six months, several hours each week, presenting the paper he was working on with Cao. In December 2005, Cao and Zhu submitted their three-hundred-plus-page paper to the *Asian Journal of Mathematics* (which I also edit), promising to give "a detailed exposition of a complete proof of the Poincaré conjecture due to Hamilton and Perelman." Their paper, which supplied many of the details that were not included in Perelman's far briefer treatment, was published in June 2006, one month after Kleiner and Lott posted their "Notes on Perelman's Papers" and a month before Morgan and Tian's paper ("Ricci Flow and the Poincaré Conjecture") went online.

I caught some flak over the fact that the Cao-Zhu paper was published about six months after it had been submitted to the journal—a period of time too short, some contended, for the paper to have been adequately peer reviewed. I take issue with that assessment, however. It's common practice throughout the publishing industry to expedite high-profile papers considered timely in nature. Furthermore, I asked several leading experts on geometric flows, including Hamilton and Perelman himself, to referee the paper, but they declined so I assumed the job as referee, having a deeper background in this subject than most of my peers. After attending more than sixty hours of Zhu's Harvard lectures, as well as carefully inspecting the manuscript and not uncovering any significant problems that could not be fixed, I recommended the publication of this paper to the rest of the board of editors. While I had not come across any outright mistakes, I could not guarantee that the proof was 100 percent correct, which is something that almost no editor can do in practice. All you can do is state, after a thorough review, that to the best of your knowledge the argument appears sound.

I then circulated the paper to all the journal's editors, asking for comments, but nobody complained or made any comments. The paper was later accepted according to the *Asian Journal*'s usual editorial procedures. I should point out that the journal's requirement of consent from the *entire* board of editors is more stringent than the policy in place at some leading mathematics periodicals.

But that didn't quell comments from critics claiming that I'd cut corners during the editorial process. I also got into trouble for a few remarks I'd made during a talk about the Poincaré conjecture on June 20, 2006, during the string theory conference I organized in Beijing. Some observers felt I had placed too much emphasis on the foundational work of Hamilton, even though Perelman himself explicitly stated, "In this paper, we carry out some details of the Hamilton program." Perelman has consistently stressed the importance of Hamilton's work, later insisting that Hamilton deserved as much credit for the Poincaré proof as he did.

Other people were particularly upset by two sentences in my roughly hour-long talk in Beijing. "In Perelman's work," I said, "many key ideas of the proofs are sketched or outlined, but complete details of the proofs are often missing. The recent paper of Cao-Zhu . . . gives the first complete and detailed account of the proof of the Poincaré conjecture." I think my first statement is unassailable. While Perelman's argument may in fact be airtight, he offered a condensed version of the proof that clearly did not furnish all the details. As for my second remark, I probably could have worded it more cautiously. There is little question in my mind that Cao and Zhu provided the first published "detailed account," but there is room for debate as to whether it or the discussions submitted by Kleiner-Lott and Morgan-Tian are "complete" or not. I never claimed, nor did I believe, that Cao and Zhu had surpassed Perelman's work in any way. Their contribution, though important, was mainly one of exposition—filling in some of the steps that Perelman's abbreviated treatment did not go into. By giving praise to Cao and Zhu, whom I believe had done the most rigorous expository work here, I was trying to encourage other Chinese researchers to be bold and take on prominent work at the frontiers of mathematics.

Promoting Chinese mathematics, which still needs a boost, is second nature to me—a tendency that was encouraged by my father, as well as by my former teacher Chern, who had made a similar commitment. That said, I don't think my encouragement was unwarranted in this instance. Cao and Zhu's effort was substantial; they made an earnest (and arguably successful) attempt at explaining the entire proof. Still, I sometimes wonder whether there might have been some backlash to my remarks—a perhaps below-the-surface sentiment that China was still a second-class citizen in the mathematics world and its representatives should therefore not try to position themselves at the forefront of the field.

Early in the Cao-Zhu paper, a few pages repeated arguments made by Kleiner and Lott, without giving them proper attribution. This was an unfortunate misstep on the part of Cao and Zhu, which occurred, they explained, when they forgot in the course of a multi-yearlong effort that their notes regarding a particular argument—that finite distance implies finite curvature—were based on the contributions of Kleiner and Lott. That lapse caused Cao and Zhu considerable embarrassment, as is appropriate even for an inadvertent error. As editor of the journal, I too came under criticism for this slip-up. A few months later, the *Asian Journal* published an erratum containing an apology from Cao and Zhu, along with an explicit acknowledgment of the prior work of Kleiner and Lott.

My reputation took a more serious hit on August 28, 2006, when an article written by Sylvia Nasar and David Gruber called "Manifold Destiny" came out in the *New Yorker* magazine. I had spent a lot of time talking with Nasar—the author of *A Beautiful Mind,* a book about the mathematician John Nash. I had much more limited contact with her coauthor, Gruber, a Columbia journalism graduate who was then studying oceanography at Rutgers. My conversations with Nasar had seemed quite amicable, and I had even helped arrange for her to go to China to attend the string theory conference I'd organized, as well as to meet with some mathematicians and physicists there. But until her article came out, I had no idea what she really had in mind.

The story Nasar crafted had a traditional structure: A hero, Perelman in her story, is driven by pure motives and untainted by the desire for wealth or fame. Pitted against this noble soul is a dastardly villain, determined to thwart him at every turn. That's the role that I got, though I never had a chance to discuss the matter with central casting.

A cartoon on the article's opening spread told all: It showed me trying to rip the Fields Medal from Perelman's neck. This illustration bothered me for all kinds of reasons. First, I already had a Fields Medal. Nobody in the eighty-plus-year history of the prize has won more than one, and I certainly was not vying for another (an impossibility anyway, given that I was well into my fifties by then, and recipients must be younger than forty in the year in which the medal is awarded). Nor was I seeking personal glory in this area. In fact, when Hamilton wanted to add my name as a coauthor of a critical paper on Ricci flow, I thanked him but said no. And in 2006, the year Perelman was granted a Fields Medal, I went on record stating that he fully deserved the award.

As Hamilton himself attested in that same year: "Far from stealing credit for Perelman's accomplishment, he [Yau] has praised Perelman's work and joined me in supporting him for the Fields Medal." Perelman, incidentally, chose not to go to the August 2006 ceremony in Madrid to accept the prize, which was given, according to the official citation, for his "contributions to geometry and his revolutionary insights into the analytical and geometric structure of the Ricci flow." The citation made no mention of the Poincaré conjecture, nor did Perelman refer to that conjecture explicitly in his 2002 and 2003 papers.

I was unhappy with other statements made in the *New Yorker* article, many of which were simply untrue. I was attacked for taking too much credit for my work on the Calabi conjecture and Calabi-Yau manifolds, even though my friend Eugenio Calabi had personally told me that, if anything, I had accorded *him* too much credit.

"Manifold Destiny" also claimed that Chern wanted to bring the International Congress of Mathematicians to Beijing but I made "an eleventh-hour effort" to move the event to Hong Kong. This was wrong on many levels. For starters, I was the one who had first proposed holding the

ICM in China. Cambridge University mathematician John Coates, who served on the executive committee of the International Mathematical Union from 1986 to 1994, confirmed that I formally wrote to the committee around 1988 about holding a future congress in Beijing. "I remember it very clearly," Coates said, "since I was asked to draft the reply."

Chern, for his part, was never that excited about the idea of hosting the congress there. And S. Y. Cheng, who was vice president of the Hong Kong Mathematical Society at the time, further maintained that I never made any attempt to move the ICM to Hong Kong. All rumors to that effect were purely fiction or, as Coates put it, "groundless gossip, as usual from unnamed sources, [as] reported in the article."

The *New Yorker* article claimed that "many mathematicians" worried that my political machinations may have damaged the mathematical profession and quoted Phillip Griffiths saying, "Politics, power, and control have no legitimate role in our community, and they threaten the integrity of our field." This is a curious assertion coming from someone who is himself a politically involved mathematician. Even Wilfried Schmid—a former graduate student of Griffiths who admitted to being deeply indebted to his advisor—recognized the irony of that claim in a letter he wrote in my defense. Nasar, Schmid noted, should have been "well aware of the extent of Griffiths' political power within the field of mathematics," where he has served as secretary of the International Mathematical Union, director of the Institute for Advanced Study, and provost of Duke University. "Unlike Griffiths," Schmid added, "Yau has never sought out influential administrative positions."

The *New Yorker* portrayed me as a bully by quoting an article from a Chinese newspaper that claimed I criticized my former student Tian for "collecting a hundred and twenty-five thousand dollars for a few months' work at a Chinese university." The magazine described Tian as "my most successful student," an assessment I vigorously reject. He was also innocently portrayed as a strict adherent to the Chinese dictum "honor your teacher," even though—as most people in our sector of mathematics would agree—there had not been a lot of honoring done, in either direction, for quite some time. I had reason to believe, moreover, that Tian was involved in a smear campaign against me, waged both in public and over the web. Multiple web attacks targeted at me and my colleagues often appeared online when I criticized his behavior. Some of these attacks contained personal information about me, including things I had told only Tian back in his student days, which further aroused my suspicions.

The main reason I had parted company with Tian is that he was the person who opened the floodgates to a practice I thoroughly disapprove of. Several Chinese universities, according to a 2006 article in *Science Magazine*, "have created 'million-yuan professorships' with stratospheric—for China—annual salaries equivalent to $125,000." The article described my former student as "the man at the center of a tempest engulfing Chinese academia. Tian is a premier example of a controversial phenomenon: a Chinese-born researcher with a full-time faculty position overseas who gets paid handsomely for short working stints in his homeland."

A million yuan was then more than ten times the typical salary of a full-time professorship at a top Chinese university. I'd heard about the cushy deal Tian was getting—receiving money that supplemented the salary he was already drawing from Princeton—long before the *Science* article came out. Other Chinese mathematicians then living abroad followed in his footsteps, accepting high-paying positions at Chinese universities while holding jobs in other countries.

I was among the first to publicly condemn this practice, which I think is wrong for many reasons—one being that Chinese professors were being grossly underpaid while graduate students often had to get by on allowances of just $50 per month; both these professors and students could have used more financial support. I've been vocal about my complaints because Tian was my student, and I wanted to make it clear that I had no tolerance for this kind of behavior.

China also bears responsibility for enabling conduct of this sort. The country initiated a program around the same time called "Thousand Talents," which spent billions of dollars to entice big-name scholars from the United States and other Western countries to go there to bolster its own universities. But China was not getting much payoff for its effort; too many visiting scholars were pocketing the money without devoting much time, or energy, to China. Indeed, the system was rife with abuse: in one case a researcher held three positions in China during the same year, while also retaining a full-time position in the United States. In the meantime, the native Chinese professors were being paid a pittance by comparison. Given that I have never taken money from the centers I've started and run in China, I don't think it's exactly fair to cast me as the bad guy in this instance, as the *New Yorker* attempted to do in its Poincaré "exposé."

Finally, the magazine alleged that my career was in decline, stating that "more than a decade had passed since Yau had proved his last major result." This contention dismissed the recent work I'd done in string theory—on the mirror conjecture, SYZ conjecture, and Strominger equations—which some people consider important, along with my efforts in general relativity and other areas. I've been pleased to hear that many of my mathematical colleagues have taken issue with that claim. The fact that I won the 2010 Wolf Prize in Mathematics—for my work in geometric analysis, which "linked partial differential equations, geometry, and mathematical physics in a fundamentally new way"—is perhaps further evidence that my standing in the field was not permanently harmed by my portrayal in the *New Yorker* four years earlier.

Still, the eighteen-page article—which seemed to be as much about vilifying me as in exalting Perelman—did not bring me any joy. While it may be difficult to read a negative review of a book you've written, it's even harder to read a negative review of the career you've led and the life you've lived, especially when the discussion is so patently biased and strewn with errors. The question before me was in figuring out how best to respond.

I discussed a possible libel suit against the magazine with a top Boston attorney. While he considered our case strong, he thought it would be a long haul, likely dragging on for a year or more. And even if I won, I wasn't sure what exactly would be accomplished in the end. Although I was upset that my reputation had been tarnished, I decided the best way to restore it would be in the classroom, or in my study, rather than in the courtroom. Instead of putting this incident behind me, a drawn-out legal battle would carry me in the opposite direction.

A story from Confucius came to mind, as often happened, for I'd been well indoctrinated by my father. This tale, which dated back to around 500 BCE, was among the first I had committed to memory. It takes place during a famine in the ancient Chinese state of Qi. A rich man, Qian Ao, handed out food with contempt to hungry people scattered along a road. A man refused to take the food from Qian Ao, even though he was starving, because Qian Ao did not treat him with respect. Qian Ao tracked down the man later and apologized to him, but the man still refused to take the food and he soon died.

The moral of this tale, as best I could glean, is that one should always insist on being accorded dignity, but at the same time one should not work against one's own self-interest by a blind, unwavering adherence to pride. I evidently learned my lesson well, as I still remember this story more than half a century later. It did not mean much to me as a child, but upon reflection I am surprised to see how much it has influenced me over the years—and how, at various instances in my life, I have come to reflect upon it.

One of those times, of course, occurred after the publication of Nasar's article with its unflattering characterization of me. But unlike the starveling in the Confucian fable above, who would not move on from the original affront, I was not going to let a wounded ego hold me back. Nor would I cling to the role of a passive victim, unfairly preyed upon by "the most prestigious magazine in America," to quote Nasar. I am a fighter by nature, disinclined to let others push me around.

While this was an unpleasant episode, to be sure, I have been through far worse moments and thus have learned to deal with the vicissitudes life throws our way. The death of my father was by far the most painful thing I've ever experienced; this was a minor irritant by comparison. Although my first instinct is to defend myself when under attack, I concluded that the best course of action in this case was to try to forget the incident and simply move forward.

Our greatest glory—to paraphrase my dad's hero, Confucius, once again—is to rise up after we've been knocked down, even, and perhaps especially, after being taken down by an unfair, below-the-belt punch. One thing that helped was a laudatory profile of me, titled "The Emperor of Math," that appeared in the *New York Times* in October 2006, less than two months after the *New Yorker* article came out. That chronology was not accidental, Nasar told me, as she was rushing her story through so that it would be published before the *Times* article went to press.

The *Times* piece was very flattering, and I believe it provided a more balanced picture of me and my work, perhaps because the author, Dennis Overbye, took his time, interviewing me off and on over the course of half a year. Nevertheless, that story, like its less complimentary predecessor, was external to me—something I had neither created nor controlled. Of far greater importance to me were things I could do on my own, by way of moving forward—things I still hoped to accomplish in my field and in my professional career as a whole. Rather than devoting further thought to articles that discussed me, favorably or unfavorably, I wanted to give my full attention to mathematics and to my research, which is something I always enjoy. I've often sought refuge in this work during times of stress, and mathematics has rarely, if ever, let me down.

I turned, for example, to a problem from general relativity that I'd been thinking about for many decades—an offshoot of my earlier work (with Rick Schoen and others) on the positive mass conjecture. The essence of this problem stems from the fact that we still don't know how to define "local mass" in Einstein's theory. We can only define the mass of an isolated system that is very far away—essentially at an infinite distance—which Schoen and I showed must be positive, for otherwise such a system would not be stable. But we'd also like to be able to describe systems closer at hand, which involves the concept of "quasi-local mass." If, for example, two black holes were interacting, what is the partial mass of that system, the mass, say, of just one black hole, rather than the total mass of the entrained pair, as viewed from afar?

These kinds of questions don't apply just to black holes, of course. Given an arbitrary closed, two-dimensional surface in space, we'd like to be able to say something about its mass too—apart from knowing that the computed value ought to be positive. In 2003, my former student Melissa Liu (now a professor at Columbia University) and I published a paper containing the first definition of quasi-local mass that was proved to be positive in all cases (except for a trivial case in which the mass could be zero). I gave talks on the subject in Cambridge, England, to Stephen Hawking and Roger Penrose—both having worked on their own definitions of quasi-local mass—as well as to Gary Gibbons and other physicists there. None of them challenged the mathematics, though they still wanted to explore the physics further. But these scholars were not shy; they surely would have given me grief if they had detected the slightest weak point in my argument.

I felt that Liu and I had made an important contribution by demonstrating a way to measure the mass and energy content of a region in space that could not be determined through previously known methods.

But I also knew this work could be strengthened. I made further progress on this score with my former student Mu-Tao Wang (also a Columbia professor) in a series of papers published from 2006 to the present day, which I believe provide the best definition of quasi-local mass so far—a definition that applies to a broader and more natural class of spaces. My work with Wang on quasi-local mass has also led to a better understanding of angular momentum and center of mass—two concepts that have remained imprecisely defined in general relativity.

In 2008, I became chair of the Harvard math department, taking over at a particularly challenging time because of the ongoing nationwide financial crisis, which threatened the collapse of the U.S. banking system. Harvard's endowment had lost $11 billion in the stock market crash and was at risk of losing billions more. Amidst fears that the whole university would fall apart, every department was asked to slash its budget, starting with 20 percent cuts across the board. I explained to the dean that the math department was already quite lean. Additional cuts to our meager budget would hurt undergraduate teaching, which is always regarded as the department's most important mission. The one area we might do some economizing, I suggested, was the $30,000 in telephone bills incurred by the faculty. I mentioned this as a gesture to show that I was not unbending. The dean soon realized there was not much savings to be reaped here, and in the end he kept our budget at its already frugal level.

The next issue was the hiring of junior faculty. Three to four assistant professors are typically hired each year, though there were questions as to whether we could afford to hire anybody this time around. I raised some money from the Simons Foundation, from a group called the Friends of Harvard Mathematics (which was headed by the French mathematician Bernard Saint-Donat, a student of David Mumford), and from other private donors, including William Randolph Hearst III, a philanthropist who graduated from Harvard in 1972 with a bachelor's degree in mathematics. With this outside assistance, we were able to make five faculty appointments in my first year, one more than usual. In 2009, we hired three high-profile professors: Mark Kisin, a promising young number theorist; Jacob Lurie, who showed remarkable abilities in algebraic geometry and category theory, among other areas; and Sophie Morel, a rising star with expertise in the Langlands program, number theory, algebraic geometry, and representation theory. I was particularly proud of getting Morel to join our faculty; in addition to being quite talented, she was also the first tenured woman in the Harvard mathematics department. (Unfortunately, she moved to Princeton three years later.)

These moves helped restore morale in the department, calming down faculty members who had been agitated over the dire fiscal situation. I raised outside money so the math faculty, undergraduates, graduate students, and postdocs could eat lunch together once every two weeks in the department's fourth-floor common area. This program, which enables people throughout the department to exchange ideas and socialize, has been a pretty big hit, but another measure has also gone far toward elevating spirits and raising productivity among our personnel: Someone had installed a very good, as well as costly, coffee machine in the department chair's room, where I was stationed at the time. I don't drink coffee, and the office assistants were annoyed with all the people lining up outside my door to fill their cups. So I moved the machine upstairs to the common space, where everyone had access to it—and that may rank as one of the most popular steps taken by any math department chair in the almost three-hundred-year history of the department. (Harvard students in the 1630s rebelled over shortages of beer but not, as far as I've heard, over shortages of coffee.)

Although I had not been eager at first to become department chair—a responsibility that rotates among the senior faculty—my term, from what I gathered, worked out pretty well. In fact, the dean was so pleased with my stewardship of the department that he asked me to stay on as chair beyond the normal three-year term. But I preferred to abide by the usual tradition, telling him that three years was enough.

In 2008 or early 2009, the president of Tsinghua University in Beijing, Gu Binglin, came to my home in Cambridge and asked me to run a math center at the university. Tsinghua had been one of China's most important universities for basic research until around 1950, when Mao decided to shift its focus toward engineering, industrial applications, and technology development. A large group of mathematicians were moved to the Academy of Sciences, Peking University, and elsewhere. Science and math programs at Tsinghua consequently experienced a marked downturn before eventually rebounding. A program in applied math began in the 1970s, and pure mathematics was reintroduced in the 1990s.

I had been approached quite a few years earlier about establishing a center at Tsinghua. It started with a letter that Gu Yuxiu, a professor of electrical engineering at the University of Pennsylvania, sent to then–Chinese president Jiang Zemin. Gu had been Jiang's professor in the 1930s at Shanghai Jiao-Tong University. In his letter, Gu argued that for China to be strong, it needed science and technology. And for China to be strong in science and technology, mathematics, and especially basic mathematics, was essential. Owing to that letter, a copy of which Gu sent to me as well, President Jiang resolved to improve mathematics at several universities, including Tsinghua.

Word was passed down to Tsinghua's president, Wang Dazhong, and Wang asked me around the year 2000 to run a math center at the university. After I realized that Wang had not succeeded in lining up any funding, I told him that without any money there was nothing I could do. And as far as I could tell, Wang did not press further.

My former student Kefeng Liu told me that Wang also approached Chern about starting a math center at Tsinghua, but when Chern asked about the math department having its own library, that request was turned down.

Things seemed noticeably different eight years later when Tsinghua's next president, Gu Binglin, visited me in Cambridge. I immediately sensed that his interest in establishing a new math center was genuine. Xi Chen—the school's "Party secretary," as well as its top decision maker—was also determined to make Tsinghua a first-rate university. Gu, a physicist, promised me the resources to make the Tsinghua center a success. We both agreed that the new center might be the catalyst needed to ramp up mathematics at the university, with a potential spillover effect throughout China. I soon went to Beijing, where Xi Chen made the same guarantees, and this time I committed to the venture.

As mentioned in the previous chapter, Michael Atiyah had notified me of the offer by the physicist C. N. Yang that would have put him (Atiyah) in charge of a new math center at Tsinghua University, even though Yang apparently had not let the university president Gu or the Party secretary Chen know of his plans. When I learned of this, I told the university officials, "If Yang wants Atiyah to do it, it's fine with me, but you shouldn't ask two people to run the center." This issue soon resolved itself, perhaps because Atiyah—based on what he told me—was never really interested in the position.

My first priority was to hire people for the math center, which again put me at odds with Yang, who was trying to recruit mathematicians for his institute too. His hiring philosophy was decidedly different from mine, as he tended to offer famous mathematicians big salaries for brief visits. That strategy had not been too successful, in my opinion, nor did I consider it especially healthy for the development of mathematics in China.

I met with university officials again, telling them I was happy to compete with the Chinese Academy of Sciences, Peking University, and other facilities in China, but my center was not going to compete with Tsinghua itself. I said that the university should have a consistent policy regarding the hiring of mathematicians, and since I was a mathematician, I should be in charge of appointments in that area. They agreed, and I soon was able to assemble a talented faculty.

Several years later the center was renamed the Yau Mathematical Sciences Center. With about forty mathematicians employed there by late 2014 on a full-time or part-time basis, I believe this center is turning out a strong cadre of Chinese professionals trained to carry out quality research in the Western style that was just beginning to take hold in the East.

Lest I give the impression that I was wholly focused on the East, I should mention that I took the lead in starting the Center for Mathematical Sciences and Applications (CMSA) at Harvard in 2014. I helped raise $200 million from the Evergrande Group, China's largest real estate developer, to found three new centers—the CMSA, the Harvard Center for Green Buildings and Cities, and the Evergrande Center for Immunologic Diseases at Harvard Medical School. Some of my peers have said, perhaps only partly in jest, that I'm better at fundraising than I am at geometry. While I'm glad to have helped some worthy causes—by starting various math institutes and related activities—when all is said and done, I'd rather be known for my work in mathematics than for my success in "shaking the money tree."

As for why I wanted to start CMSA in particular, I was motivated for a couple of reasons. I've long felt that while Harvard has one of the best departments in the world as far as pure mathematics goes, if anything, it is "too pure." People in the department have a strong aversion to applications, and that attitude is hard to shake. Getting the department to take on people in applied math and interdisciplinary subjects, as I've tried to do, is difficult because different criteria should come into play when appraising candidates, though those charged with the hiring decisions tend to apply the usual standards. The great mathematician David Mumford moved to Brown University in 1996 because he wanted to do more applied work and did not feel he could get the support he needed at Harvard.

Given that we live in a modern world where mathematics is becoming increasingly important in so many different areas—including biology, chemistry, economics, engineering, and, of course, physics—I feel that we can no longer afford to ignore applied math altogether. It's also true that many illustrious mathematicians of the past—people like Euler, Gauss, Riemann, Poincaré, and Hilbert—did not consider themselves above dabbling in applied areas. With that in mind, I took the initiative to raise outside money to create new opportunities in applied mathematics—a move that was strongly embraced by Harvard's dean and provost, who were eager to see more interdisciplinary research carried out at the university. My specific hope was for CMSA to fill the gaps in areas that the math department was not covering. With help from my Harvard math colleagues—including Michael Hopkins, Clifford Taubes, and Horng-Tzer Yau—as well as the applied mathematician and physicist Michael Brenner and the statistician Jun Liu, I think we're off to a good start.

It's true that most of the work I'm known for is of the "pure" mathematical variety, but I have delved, upon occasion, into "impure" terrain as well. In the early 1990s, I did some work in graph theory—the study of graphs that can shed light on processes in various physical, biological, and social systems—with Fan Chung of the University of California, San Diego. I've collaborated with my brother Stephen on papers in nonlinear control theory—an area of applied math that has been used extensively in industry. When Mumford left for Brown, I inherited his computer science PhD student David Gu. Gu and I, along with other students and postdocs, have since been applying some of the tools I developed in the Calabi conjecture proof—involving conformal geometry and Monge-Ampère equations—to computer graphics, which has been used, in turn, to do some cool new things in brain imaging and in medical imaging more generally.

I enjoy this sort of work, which for me represents a refreshing change of pace, and I'm happy to have opened up avenues for more applied and interdisciplinary math-related research at Harvard. But that's still somewhat of a sideline for me. Pure math has been—and shall remain for the foreseeable future—my main pursuit, the thing that really gets me excited. Even though I sometimes have the inclination to broaden out, I still feel that the best, and most essential, part of mathematics is pure and foundational. And that's what I've poured myself into in the wake of the *New Yorker* fiasco. I became heavily involved in a project related to string theory that I'd first gotten going on in the mid-1990s.

Mirror symmetry had provided clues regarding the close ties between string theory and enumerative geometry, a branch of algebraic geometry. I was hoping to connect string theory with number theory as well and had reason to believe these efforts would pay off. Part of my confidence stemmed from the fact that Calabi-Yau manifolds stand at the center of string theory (see Chapter 8). A one-dimensional Calabi-Yau is called an "elliptic curve," and the theory of elliptic curves is in turn one of the deepest subjects in mathematics, lying at the heart of number theory. Given that the generalization of elliptic curves to higher dimensions is Calabi-Yau manifolds themselves, I suspected that a thorough understanding of these manifolds might help bring physics (in the guise of string theory) into number theory and number theory into physics. At least it was not so farfetched to think this could be a productive avenue.

Eric Zaslow and I made a start on this front in a 1996 paper in which we counted the number of "rational" curves on a K3 surface—a two-dimensional Calabi-Yau space, as well as a complex, two-dimensional elliptic curve—showing that this number, an integer, is related to the eta function, an important formulation in number theory that was introduced by Richard Dedekind in 1877. But our analysis concerned only a limited class of curves—those of genus 0, which loosely refers to curves (or surfaces) with no holes. The Harvard physicist Cumrun Vafa and three other collaborators made an important contribution toward the problem of counting curves of higher genus in a three-dimensional Calabi-Yau manifold. I built upon this work in 2004 with my then-postdoc Satoshi Yamaguchi, providing some new ideas regarding the structure of the counting function. I'm still pursuing the topic with other people, as I believe this counting function may eventually prove to be a generalized form of the eta function. Firming up this connection could in turn strengthen the link between string theory and number theory, while also paving the way toward various applications in number theory.

In related work involving tie-ins between string theory and number theory, Bong Lian and I proved that the number of curves on a quintic Calabi-Yau space is divisible by 125 in certain cases—those in which the degree of the equation defining that curve is not divisible by 5—thus answering a question posed by the algebraic geometer Herbert Clemens. Over the past decade or so, Lian and I also have been building upon ideas inspired by mirror symmetry in order to calculate the period of integrals, which relates to a problem, still not fully solved, discussed all the way back in the 1700s by Euler.

I feel confident that string theory can lead us to important new avenues in number theory, although this is a long-term effort in which we've so far barely scratched the surface. The key advances may well be made by others, rather than by me, and I'm fine with that, intent for now on trying to get the ball rolling in this direction.

As I've said before, I always like to have a variety of problems to work on or think about in my spare time, whether I'm driving or sitting in the dentist's office. And there are, at the moment, quite a few other items on my plate, including the aforementioned Strominger equations from string theory, which could shed light on the expansive and largely murky domain of non-Kähler manifolds. The best advances in mathematics, in my opinion, are not those that solve problems, thereby closing off an area of research, but rather those that open up a whole new range of problems and related issues to explore.

One problem I'm not actively working on is the Poincaré conjecture, as I'm happy to put the controversy surrounding it behind me. But I can't keep my mind from turning to that problem, upon occasion, and I still have some lingering doubts that—if expressed out loud—are likely to get me in trouble. Although it may be heresy for me to say this, I am not certain that the proof is totally nailed down. I am convinced, as I've said many times before, that Perelman did brilliant work regarding the formation and structure of singularities in three-dimensional spaces—work that was indeed worthy of the Fields Medal he was awarded (but chose not to accept). Perelman built upon a foundation painstakingly laid down by Hamilton and carried us further along the path laid out by

Poincaré than we've ever ventured before. About this I have no doubts, and for that, Perelman deserves tremendous credit. Yet, I still wonder how far his work involving Ricci flow "technology" has taken us. And I also can't keep from wondering whether another approach—making use of some of the minimal surface techniques I developed many years ago with Bill Meeks, Rich Schoen, and Leon Simon—might lend some clarity to the situation.

In 2003, Perelman told Dana Mackenzie, a reporter for *Science* magazine, that it would be "premature" to make a public announcement regarding a proof of the geometrization and Poincaré conjectures until other experts in the field weighed in on the matter. Confirmation of this proof resided largely with outside "experts," given that Perelman receded almost completely from the mathematics scene, which is a great loss to the field. The thing is, there are very few experts in the area of Ricci flow, and I have not yet met anyone who claims to have a complete understanding of the last, most difficult part of Perelman's proof.

In 2006 or thereabouts, a visiting mathematician who was knowledgeable about this area stopped by my Harvard office to reproach me for raising questions about Perelman's work. Yet he admitted, when I asked him, that he did not entirely grasp the latter part of Perelman's argument. That's no knock on him, as that admission puts him in a rather sizable group. In fact, I don't know whether anyone else, including Hamilton, has fully gotten it, and I'd put myself in that category as well. As far as I'm aware, no one has taken some of the techniques Perelman introduced toward the end of his paper and successfully used them to solve any other significant problem. This suggests to me that other mathematicians don't yet have full command of this work and its methodologies either.

Hamilton, who's now in his seventies, has told me that it is still his dream to prove the Poincaré conjecture. That does not mean that he thinks Perelman did anything wrong. Hamilton, a truly independent spirit, is not one to follow in someone else's footsteps, nor would he be inclined to "connect the dots" of another's argument. He just may want to do it his own way and complete his life's work of the past three and a half decades.

Nevertheless, that still leaves me with the sense that this situation is not unequivocally resolved, perhaps leaving theorems of incredibly broad sweep hanging in the balance. Expressing any doubts on this subject, I know from experience, is a politically fraught proposition. But for the sake of my own questions—and for mathematics as whole—I'd still like to be more certain of where we stand. If that makes me a pariah, so be it. In the end I care more about mathematics—the path I chose to follow more than a half century ago—than I do about what others think of me.

CHAPTER TWELVE

Between Two Cultures

WHEN I FIRST SET FOOT in the United States in 1969, as a twenty-year-old who'd never ventured far from home, I was struck more than anything by the stunningly blue sky—its clarity and crispness providing hope that I might soon be able to see farther, and that perhaps the secrets of mathematics would someday be revealed to me.

My reaction upon arriving in China a decade later—my first visit to the country since I was an infant, oblivious to his surroundings—was much more visceral and instinctual. I bent down to touch the soil, as if trying to make a connection with the ground from which my ancestors had arisen and I had later sprung forth. Acting on impulse, I was reaching out to the land I'd heard so much about, and was such an important part of my being, but had never before spent a moment in as a self-aware individual. While I'm not normally given to big displays of emotion—and am known, if anything, for my stoic bearing—that experience left me shaken.

Flash forward to the present: I now regularly travel back and forth between China and the United States at least a few times each year. The experience has become almost routine, though with each visit I learn something more about the two places I feel most at home in, though never fully so. I'm no social critic and am not referring here to any grand insights into these two profoundly different cultures. I just might notice some small quirks or minor annoyances that distinguish the two milieus in ways that hadn't registered before.

Some parts of my daily regimen are the same no matter where I happen to be, whereas other parts of my life can be utterly different. At the time of this writing, in the fall of 2017, I find myself in Beijing, spending a sabbatical year at the Mathematical Sciences Center of Tsinghua University that bears my name (which could prove helpful, I suppose, in the event of a momentary identity lapse, as there are plenty of reminders around). While I've never been a coffee drinker, I do like to start my day with a strong cup of tea, preferably of the Chinese variety, of which I always keep an ample supply on hand, and that's pretty much a constant wherever I go. My general approach to mathematics is another thing that does not change, regardless of whether I'm operating on China Standard Time or back at Harvard on Eastern Standard Time or somewhere in between, such as Greenwich Mean Time.

But on a practical level, I do notice appreciable differences between the two countries. For one thing, I have many more colleagues I can work with in the U.S. than in China. Collaborations have always been important to my research, and in America I've been able to team up with brilliant mathematicians from all over the world—a situation that would be hard to match anywhere else. In China, there's a much smaller pool of people with whom I can have mutually beneficial exchanges of ideas. Also in China, Google searches are essentially banned, and e-mail communications are sometimes restricted—nuisances, for sure, though nothing that changes my life and work habits in a fundamental way.

Dealing with the university administrators, however, is totally different in America than in China. If I make a request of some sort at Harvard, for instance, I typically get a note from a dean that spells out in unequivocal terms what I can or cannot do. There's room for subsequent discussion, and perhaps clarification, but the process tends to be quite straightforward. In China, the opposite is often true.

As an example, my former graduate student Kefeng Liu filled me in on his visit in the early 1990s with S. S. Chern at the Chern Institute of Mathematics at Nankai University. Liu had gone to Nankai not long after Chern, the institute's founding director, had stepped down from his post. A new director was hired in 1992, but Chern still had a big role in overseeing the center. I asked Liu how Chern was doing. "He's fine," Liu said, "but unhappy." Liu found this perplexing because the new director was doing everything Chern had asked for. Liu's confusion stemmed from the fact that he didn't understand the Chinese way of doing things, which is perhaps best described as byzantine: Venerated scholars, highly respected by the government, may say one thing publicly when they are really, and secretly, pushing for another outcome, which they don't feel comfortable saying out loud until others make the case for them.

While the new director thought he was complying with Chern's wishes, he was actually doing the opposite of what Chern wanted—hence Chern's gloomy disposition. This story line follows the old-fashioned Chinese approach, which I consider to be a peculiar and rather circuitous way of doing business. Chern, as far as I could tell, didn't think there was anything wrong with this system; he appeared to accept the status quo. Liu, however, had no clue as to what was going on beneath the surface.

When I talk to university administrators in China in my capacity as the director of several math centers, I usually find them to be quite polite. The people I deal with—deans, department chairs, university presidents, and so forth—tend to promise all sorts of things, in conversation though not in writing. But when it comes time to carry out those pledges, they often are unable, or unwilling, to follow through.

Tsinghua University, where I'm presently stationed, is somewhat of an exception—and better in this respect than most Chinese universities—having adopted a management style that more closely follows the Western approach. Nevertheless, the academic system in China is more complicated because major universities are under the control of the government through the Ministry of Education. Leadership changes at universities, which happen periodically, can result in significant upheaval. When new people come in, they don't want to do what their predecessors agreed to because in that case the successors won't get much credit. They want to have something new to show their superiors, which means doing something different, even if that means curtailing a successful program and replacing it with an ineffectual one. This introduces an element of uncertainty to operations in Chinese universities that does not exist in their U.S. counterparts.

Every university in the United States, to be sure, has its own internal politics—the inevitable squabbles within departments, between departments, and between the faculty and administration. But when the country as a whole elects a new president, that doesn't usually affect anything at the campus level—unless, of course, major funding cuts or policy shifts are instituted as a result of a change at the top.

Given the closer ties to government found in Chinese institutions of higher learning, Chinese academics have far greater incentives to move up the ladder of political authority. Within this hierarchy, all members of the university administration have a ranked level attached to them. You'll get a more generous salary and preferential treatment at, say, a hospital or an airport the higher the number you've been assigned.

Ding Shisun, the former mathematics chair at Peking University, went on to become president of the university and then chairman of the China Democratic League, one of the country's eight "democratic" parties. Ding, accordingly, has wielded a lot of power in his career. He used his influence, among other things, to aid the rise to power of my former student Gang Tian, who now holds a part-time professorship at Peking University from which he received a master's degree before earning a PhD at Harvard under my direction. Tian is also a high-ranking official in an advisory body known as the Chinese People's Political Consultative Conference, and he recently became a vice president of Peking University, which places him at the deputy minister level within the Chinese establishment. With this appointment, Tian became a potent force at Peking University, poised to take over the reins from Ding, who is now ninety years old.

Tian let me know early on that his ambitions went well beyond mathematics. He told me in 2001, as we sat in the Boston Common, that he someday hoped be a leader in China, eventually becoming one of the most powerful people in the country. I was content with a career in mathematics, but I try to stand by my students and help them with their careers, even if the choices they make and paths they choose don't always accord with mine.

I wish that relations between Tian and me hadn't deteriorated so badly, and ideally I wish we could be on better terms. But before a full reconciliation can happen, I would like to see him make amends for practices that strike me as improper. Simon Donaldson and his colleagues, for example, have accused Tian of appropriating their ideas without providing suitable attribution to their prior work. It seems to me that advances in Tian's career may have been hastened by behavior on his part that I consider questionable, especially in China where academic standards have not always been so rigorous as in the West.

In the United States, progress in academia is largely based on scholarly work—how good you are at what you do. But in China, political gravitas plays a bigger role, which has prompted many academics—including those in the realm of mathematics—to give short shrift to research, focusing instead on more direct means of moving up in the pecking order. And the surest path to power is by becoming an "academician"—the highest academic title in the land, a lifelong honor conferred by the Chinese Academy of Sciences to about 750 scholars in science and math and about 850 members of the Academy of Engineering.

The U.S. counterpart, the National Academy of Sciences, was established in 1863, long before the 1949 founding of the Chinese Academy. The NAS currently has about twenty-three hundred members, and I've been a member myself for the past twenty-five years. While it's definitely an honor, the NAS designation has a relatively minor bearing on one's life in a material sense. Not so in China, where academicians get many of the same perks granted to high-ranking Communist Party officials, such as private rooms at hospitals and access to VIP lounges at airports. If you're important enough to be called a "leader of China," entire rail cars can be set aside for you, and you'll be rewarded with a more comfortable income as well. Beyond these and other individual benefits, there are more far-reaching effects. In China, most people will agree with something if enough academicians say it's true. If you don't have any academicians on your research panel, that panel won't carry much weight in the eyes of the government. If, on the other hand, three or more academicians write a joint letter to the government, that letter is likely to land on the desk of the premier.

Just as a university's stature depends on how many academicians are on its faculty, a city's measure of academic excellence depends on how many academicians live there. A faraway province in Tibet might have only two or three academicians; a request made by one of them has to be taken seriously because if he or she threatens to leave, the province as a whole will have diminished status. Therefore, almost nobody dares to offend an academician. They are treated like royalty, without necessarily having done much to earn their lofty titles (and in that respect they might be like royalty too).

The people in charge of the Chinese academic world have rather limited judgment regarding those who excel in their field, or not, and they often refuse to seek advice from outside authorities. There are not enough experts in China to do the necessary assessments, and many of those with expertise can be bought off with favors offered by the applicants and their sponsors. This can lead to some dubious choices regarding who gets into the academy.

Because I do not reside in China, nor do I have a Chinese passport, I don't have the right to vote for new members of the Chinese Academy. But I was consulted around twenty years ago regarding the candidacy of a Chinese mathematician who worked in the field of dynamical systems. This candidate, incidentally, happened to be the brother-in-law of a senior member of the academy. He also had the enthusiastic support of another dynamicist, a Chinese-American mathematician, who asked me to help push his friend's appointment through.

I wasn't familiar with the candidate's work, so I asked some of the world's top experts in dynamical systems—including Michael Herman and John Mather—as to how good they thought he was. Their verdict was that his work was only about average, even in China. I passed these letters on to the Chinese Academy president, who in turn brought the matter up for debate by the election committee. The aforementioned senior academy member, who participated in this discussion, argued that the letters should be disregarded because the writers were not Chinese. These matters, he insisted, should be handled by Chinese people, and Chinese people alone. His view prevailed. The candidate (his brother-in-law) was duly elected an academician and, a few years later, became president of the Chinese Mathematical Society.

Because of the prevalence of lobbying—and a tendency among members to vote for their friends, allies, family members, and anyone else they think it might be advantageous to support—intellectual prowess is often a secondary factor when it comes to the selection of academicians in China. As a result, too many academy members care little about research; their primary concern seems to be pursuing personal advancement by making other people happy.

This arrangement is far from ideal. Many observers believe, as do I, that one of the major obstacles for the development of science in China is the academicians themselves—the people who, above all others, are supposed to be the exemplars of scholarly achievement throughout the land.

Since my primary residence is in the United States, I am not eligible to be an academician, not that I ever tried or cared to become one. However, I was named a foreign associate of the academy in 1995, which was when the category was introduced. C. N. Yang became a foreign associate at the same time, as he was then living in the United States, holding the Albert Einstein Chair at Stony Brook University.

Tian became an academician in 2001. He had tried to gain admittance earlier but was not eligible because of his full-time academic appointment in the United States. After Tian promised to return to China on a full-time basis—a vow that would have been difficult to fulfill, given his faculty position at Princeton (which was later reduced to half time)—his admission to the Chinese Academy was formally taken up. The discussions about candidates, including Tian's case, went on for several days, and Tian did not have a clear majority. So K. C. Chang, who supervised Tian's master's degree work at Peking University, along with one or two other academicians, violated the normal protocol by dispatching someone to the home of a member who was too ill to be at the meeting. This member (who had a high fever) was then brought to the meeting just long enough to vote for Tian, although he should not have been allowed to cast a ballot because, according to the academy's rules, he first would have needed to attend most of the prior discussions. Yet it was by virtue of that single vote—made as a result of a last-minute rousting and some rule bending—that Tian got into the Chinese Academy.

About a decade earlier, Chern was pushing for his Berkeley associate Wu-Yi Hsiang to become an academician at the prestigious Academia Sinica (a predecessor to the Chinese Academy of Sciences, which moved to Taiwan in 1949, just before the Communist takeover). In 1991, Hsiang claimed to have proved a famous problem posed 380 years earlier by the astronomer Johannes Kepler. Kepler's conjecture, also called the "sphere packing problem," concerns the densest way to stack round objects (or spheres) in a square box. If those round objects happened to be oranges, all of the same size, the question comes down to this: What arrangement would allow you to fit the most oranges into the box? The optimal configuration, Kepler posited, would be one in which each orange sits in the hole formed by three oranges below it, with each orange in the middle of the box (though not along the edges) touching six others in total. David Hilbert repeated the question in 1900 with slightly modified wording, labeling it number eighteen on his widely heralded list of unsolved mathematical problems.

This was the problem Hsiang alleged to have cracked—a challenge, he said, that led him to develop a host of new tools in spherical geometry. His paper, "On the Sphere Packing Problem and the Proof of Kepler's Conjecture," appeared in the October 1993 issue of the *International Journal of Mathematics*. Chern felt that electing Hsiang to the Chinese Academy would be a just reward for this achievement. He vigorously recommended Hsiang's admittance in a meeting with other academicians; Wu-Chung Hsiang also spoke enthusiastically about the candidacy of his younger brother.

Some participants in that discussion, who still needed further convincing, sought my opinion. I recommended a more cautious stance that relied on the views of outside experts in addition to the endorsements of close friends and family members. As it turned out, the top authorities on the subject, John Conway (of Princeton), Thomas Hales (then at the University of Michigan), and Neil Sloane (then at the AT&T Shannon Laboratory) all found Hsiang's argument to be unsound—containing "serious flaws," according to Conway and Sloane, and having "major gaps and errors," according to Hales. In view of these statements, I said it would be hard to support Hsiang's entrance to the academy on the basis of his work on this problem; Hsiang did not get in when the vote was subsequently held.

A month or so later, when I was visiting the Chinese University of Hong Kong, C. N. Yang called me into his office. "You have offended your teacher, Chern," he said, because my comments had gone against Chern's wishes. I replied that I didn't say anything until asked and then felt obliged to answer honestly. "All you had to say was that the proof was right!" countered Yang, before promptly kicking me out of his office.

And that provides a clue as to how some Chinese academics behave: While I believe that the truth in mathematics is not subject to our personal will or ambitions—that it is part of the natural order and therefore inviolable—others may take a different view, one in which expedience has a place and can, if necessary, trump scientific fact.

In 2017, at the age of ninety-four, Yang became a full-fledged academician within the Chinese Academy where he previously had been just a foreign member. His appointment was big news in China, and with this new title he became more influential in Chinese academic circles than ever.

That influence, of course, was based on real substance—some very important achievements in physics. The ideas Yang developed with Robert Mills, generalizing the fundamental work of Hermann Weyl from the late 1920s, culminated in the "Yang-Mills theory," which occupies a central place in the Standard Model of particle physics. The Standard Model, in turn, successfully encapsulates our current knowledge of the observed universe, describing all the particles that have been seen in nature and the interactions between them. Ironically, Yang has expressed reservations about some critical aspects of that overarching theoretical framework and never seemed fully comfortable with it. Even so, his accomplishments with Mills, along with his separate Nobel Prize–winning work with Lee, were still towering feats, and particle physics has benefited greatly in their wake.

For some reason, Yang was motivated to write a letter in 2003 to the chair of the physics department at Tsinghua University, BangFen Zhu, recommending that "there must not be any additions of new faculty in particle physics [or] nuclear physics. Existing faculty in these areas should be encouraged to change to other fields." The reason Yang cited for this policy edict was that his field was "dying," although others might counter that Yang—whose famous work with Mills had taken place a half century earlier—had long been out of touch with developments in the field. In 2012, nine years after Yang submitted this letter, the Higgs boson was discovered, constituting a monumental finding in particle physics. In the same year, a new kind of neutrino oscillation was discovered in a Chinese laboratory, which held major implications for why the universe is dominated by matter rather than by antimatter. These and other accomplishments suggest that reports regarding the death of particle physics were, to quote the humorist Mark Twain, "greatly exaggerated."

In 2016, Yang wrote an article titled "China Should Not Build a Super-Collider Now." A large group of prominent physicists from China, the United States, Europe, and elsewhere had been enthusiastically calling for the construction in China of the world's largest and most powerful particle collider—a machine intended to serve as a successor to the Large Hadron Collider near Geneva, Switzerland. I played an active role in backing this effort because I believe the project would be good for China, good for physics, good for international relations, and even good for mathematics, as breakthroughs in fundamental physics have long served as a fruitful source of ideas for mathematicians. The converse is also true, and it's fair to say that both fields have benefited from this cross-pollination.

Yang, however, dismissed this whole endeavor—aimed at understanding the universe on the smallest, most basic scales—as "a bottomless money sink." He even went so far as to force the cancellation of a November 2016 lecture scheduled to be delivered at Tsinghua University by Yifang Wang, the director of the Institute of High Energy Physics at the Chinese Academy of Sciences, who was spearheading the Chinese collider project. Yang was able to pull the plug on this lecture at the last minute, even after posters publicizing the event had been plastered all over the Tsinghua campus and elsewhere in town. Wang instead gave a well-attended public lecture on the collider—which is currently in the research and development phase—at Peking University in December 2016.

I believe that Yang's motives are good and that he's genuinely trying to promote physics as he sees fit, but I also feel that a person in his midnineties, far removed from active research in his field, should not hold so much sway over other, younger physicists and over scientific research in general. This strikes me as a manifestation of an endemic problem in Chinese science and society at large: Despite gains made in recent decades by young researchers, the oldest people still hold the most power—a fact that's especially true among academicians.

There is, of course, a rather durable historical tradition at play here. The Chinese dictum "respect your elders" dates back at least twenty-five-hundred years to the time of Confucius. This attitude is formalized in the doctrine of "filial piety," which considers it an obligation and a virtue to honor one's parents, elders, and ancestors. I, too, subscribe to this notion, which is deeply entrenched in Chinese culture. I've always tried to live my life in a way that both my mother and father would approve of, and by and large, I think I've done all right on that score.

Nevertheless, that same attitude can, and frequently does, go too far, to the detriment of society at large. In the United States, most people over seventy do not hold much sway in the academic world. But not in China, where "the older the better" tends to be the rule.

Yang, to take one example, is clearly a scientist of the first rank, a commanding figure in his field. In addition to contributing so much to physics directly, the 1957 Nobel Prize that he shared with T. D. Lee gave confidence throughout China that even people from that country, which lagged far behind the United States, European nations, and Japan, could still achieve greatness on the worldwide stage. The importance of that feat cannot be overestimated. But it's also clear to me that it's long past time for people of his generation to loosen their grip on Chinese science so that younger researchers have a chance to step up and make their own marks.

I have enormous respect for Chern, too. He was unquestionably a great mathematician who made huge contributions to geometry. He built up the math departments at Chicago and Berkeley and founded MSRI, while furthering the careers of so many people, including me. For that I am forever grateful. I'm also deeply disappointed that we never managed to heal the rift between us. But in the late 1970s—about a decade after I first arrived at Berkeley with his help—I felt the need to go my own way, and I don't think Chern ever forgave me for that. And those ill feelings toward me were not limited to Chern. In China, almost everyone holds it against you if they think you're rebelling against your teacher—even if you're not rebelling at all, but just trying to assert yourself and realize your own goals.

To me, there's little doubt that the research culture in China has been held back by the dominance of the old guard, steeped in traditional ways of doing things that are anachronistic in the modern era. And this problem has only been compounded by the corrupting influence of the academicians.

Does this mean the situation is hopeless? I don't think so, or otherwise I would not be spending so much time running a half dozen mathematics centers in mainland China, Hong Kong, and Taiwan, nor would I devote myself to other math and science causes there. Because in the end I believe change will come and that it cannot be held back. In a word, I'm betting on youth. To me, it seems inevitable, as well as natural, that young leaders in math and science, who bring fresh perspectives to their fields, will emerge and gradually gain influence, in time having a transformative effect on academia as a whole.

I'm trying to foster this process by doing things differently, and establishing a true meritocracy, at the Chinese centers I run. And we should be able to do that as long as we can keep the funding going, which is one reason I continue to raise money from private donors. These centers are mainly populated with younger mathematicians—who are well before the age when one would ordinarily think about becoming an academician—and I'm helping them appreciate the rewards that come from doing excellent work, divorced from any political concerns.

That appears to be the case at the Tsinghua center, where we've assembled a large and capable group that is turning out high-quality research. If we can maintain a critical mass of people who share this work ethic, we might be able to establish a foothold in China that serves as an example for other math and science institutes. But it will be a struggle. When we have tried to draw attention to the work of our young scholars at the Tsinghua mathematics center, a few leaders at Peking University's school of mathematics seemed determined to suppress that well-earned recognition.

Although there will always be those whose main concern is money and power, I suspect that a growing number of youthful researchers have come to view academic achievement as the most important thing. And that could be the future of mathematics in China as others come to adopt this attitude as well.

Rather than focusing solely on university students, postdoctoral researchers, and junior professors, I'm also helping Chinese high school students get a taste of real research by initiating in 2008 the High School Mathematics Award. The program is modeled after the U.S.-based Science Talent Search first sponsored by the Westinghouse Corporation and then by the Intel Corporation and Regeneron Pharmaceuticals. The idea was not to have students compete to solve the standard problems presented in the annual Math Olympiad contests but rather to encourage creativity and collaboration by allowing students the freedom to work on problems of their own choosing that will take time, effort, and ingenuity to solve.

These competitions are part of a broader attempt on my part to counteract years of education in a rigid system in which Chinese students are trained to memorize things—to be passive receptacles that do whatever their teacher says. But true research is something else altogether. It's not just solving the problems your teacher gives to you but moving ahead of your teacher, at least in the particular area you're investigating.

I have no doubts that Chinese students can become more inventive, like their American peers, if they are given the encouragement and space to think independently. That's what these competitions are all about, with winners picked on the basis of their creativity as well as proficiency.

In 2013, the High School Physics Awards were launched in China, and in 2016, Science Awards were introduced in biology and chemistry. Every year, distinguished scientists—physicists like Nima Arkani-Hamed, Brian Greene, and David Gross (a Nobel laureate), and mathematicians like John Coates and Terry Tao (a Fields medalist)—have traveled to China to serve as judges.

About two thousand students from 850 teams and three hundred schools compete in a typical year. In 2015, for example, one-third of the twenty-four mathematics award winners were admitted to elite colleges abroad. Not so long ago, I might have doubted whether any of those students would return to China after completing their studies, but that situation has changed. Thanks to a rapidly expanding national economy, which has sustained annual growth rates above 10 percent for the better part of three decades, salaries in China are becoming increasingly competitive. As a result, I've had an easier time getting talented people to work in my math centers, and I believe this trend applies nationwide.

Despite the problems I have catalogued regarding China's higher educational system, in some ways the country is doing better than the United States. For example, in recent decades, the U.S. has spent trillions of dollars on wars in Afghanistan and Iraq, posing a huge drain to the economy. Meanwhile, research and development funding in science and math has languished. China, however, has generally managed to steer clear of such prolonged and costly military engagements. This has made more money available for domestic spending—to build up the infrastructure, raise living standards, and boost funding in science and technology research. While U.S. universities are still vastly superior, each side has something to learn from the other.

I try to take the best of both cultures, approaching some problems from a Western vantage point and others from an Eastern perspective. I've been heavily influenced by Chinese culture and find myself becoming fonder of reading Chinese literature and history than ever before. I even write poetry (in Chinese characters) from time to time to express my feelings or frustrations, or simply to relax.

This grounding in Chinese tradition and pastimes, which seems to be an inextricable part of my being, makes me different from my American peers. Yet it's also a fact that I've been in the United States for almost fifty years, which makes me different from my Chinese peers as well. The best parts of Chinese culture were passed on to me by my father—who taught me Confucianism, Taoism, and his personal code of ethics—and by my mother. Yu-Yun and I, in turn, have tried to pass on some of these ideas and values to our sons, who I'm pleased to see have grown up to become nice, as well as accomplished, young men with families of their own.

Although "respect your elders" can go too far, and pose unnecessary obstacles to younger generations, it clearly can be a positive principle as well. Chinese children are trained to stay loyal to their family and friends. Rather than being cast aside, older people are more closely integrated into society and therefore made to feel more secure than is often the case in the West, where elderly folks are sometimes left to fend for themselves. I suppose this could become a source of comfort for me one day, in the not-so-distant future, when I enter my so-called golden years.

In my experience, people in China tend to pay more attention to history, which has its good and not so good points. During the Qing Dynasty, roughly from 1600 to 1900, little mathematics was done in China because most scholars instead devoted themselves to the history of mathematics. Of course, there is great value in studying mathematical history, so that you can know what your predecessors (for me, geometers like Gauss and Riemann) did. I've found that perspective very helpful, whereas many Americans I know aren't inclined to look back that much. After spending a long time working on a problem, they tend to be surprised when I tell them from where, and from whom, the original ideas came.

Another thing I appreciate about traditional Chinese philosophy is that we, as people, tend to see ourselves as being part of nature, which implies that it is not in our interest to try to conquer nature. Americans don't always adhere to this view, and the intent in the U.S. often seems to be to analyze nature in order to control it. In these modern times, the Chinese don't always adhere to this view either, though it is, at least, a long-held tenet within the culture. The best course, to me, seems to lie in a mixture of these two perspectives: We can try to understand nature, which is a valuable pursuit in its own right, while at the same time trying to go along with it—to coexist and be part of the "oneness" that's sometimes referred to as the Tao.

I often wonder why China has not produced scientists of the same stature—and in the same numbers—as Western culture has turned out. I have been able to accomplish more than most Chinese mathematicians, perhaps because of the historical and philosophical grounding I got from my father, combined with spending more time in the United States, where some free-spirited American ways have, no doubt, rubbed off on me.

I have a lot of gratitude toward America, where I've been treated quite well for nearly half a century. The U.S. math world in particular has been very inviting. A lot of effort is put into nurturing young scholars, which is something I appreciate. Moreover, researchers from all over the world are made to feel welcome here. As a result, I have been exposed to a broad diversity of ideas that has in turn greatly contributed to my way of thinking about mathematics. I feel that I can be more outspoken in America, which is not always possible in China, where one has to watch one's words more carefully. Students and colleagues have, for the most part, been extremely tolerant of my hard-to-understand accent. I also appreciate the fact that if you do well in your field in America, you can be reasonably confident of advancing, whereas in China, individual performance is not always enough to get you moving forward.

That said, I still feel very strongly about China. In particular, I am committed to changing the climate in which education and research are carried out there. Things have been improving on this front, especially in recent years, and I'm glad that I might have had a hand in that.

So where does this leave me? Despite my passion for China, and the deep-seated drive I feel to help further its progress, the fact remains that I spend most days of a typical year in the United States, to which I don't have nearly as strong an emotional attachment. Yet that's where my children were born and raised; that's where our family home is; and that's where my full-time job lies. Which leaves me, as I said, in the rather peculiar place of never feeling completely at home in either America or China. My true home seems to lie somewhere in between (perhaps somewhere around the International Date Line that zigzags through the middle of the Pacific Ocean). The one thing that has made it easiest for me to move, unencumbered, between these two countries and cultures—and, indeed, anywhere else in the world—has been mathematics itself, which has long served as my true passport.

I've had a long run in mathematics, it being almost fifty years since I arrived as a graduate student at Berkeley, but I'm not yet ready to hang up my straightedge and compass. There are quite a few problems that I've started and still intend to pursue, as well as others I haven't gotten around to yet but am keeping on my "to do" list.

On the other hand, I don't want to overstay my welcome, issuing proofs far into my dotage that are not up to standards and will only cause discomfort among my colleagues and friends. When I can no longer contribute through research, I intend to focus on teaching. Seventy students have earned their PhDs under my supervision so far, with several more in the pipeline. Hamilton once said that I'd built up "an assembly of talent . . . brought together to work on the hardest problems." I hope he's right, although I'm proud in any event, because what these folks have done so far, and will do into the future, surely eclipses anything that I could accomplish as an individual. Nevertheless, a time will come when I can no longer contribute through teaching either. At that point, I intend to step aside—and I hope to do so willingly—lest I become part of the old guard I've been fighting against for so long.

But I'll always be thankful that, after much tumult in my early years, I was able to find my way to the field of mathematics, which still has the power to sweep me off my feet like a surging river. I've had the opportunity to travel upon this river—at times even clearing an obstruction or two from a small tributary so that the waters can flow to new places that had never been accessed before. I plan to continue my explorations a bit more and then, perhaps, do some observing—or cheerleading—from the riverbanks, a few steps removed.

It's been an eventful journey so far—or at least eventful for me—though I hope others have found something of interest in this rambling account of a poor boy from Shantou who stumbled upon a quest for some deep truths of nature and has maybe been lucky enough to catch a few glimmers of insight along the way.

Epilogue

BEFORE THE START of a big experiment or research program, participating scientists often talk about the things they hope to learn from the exercise, adding that there are always surprises in endeavors of this sort. Some may be in the form of frustrating setbacks, while others may be unexpected treats. Life, I've found, can be like this too. You can map out an entire week on your planning guide, listing the activities you'll be engaged in at each hour, but every day something happens that is not on your agenda, including pleasant surprises.

In early 2018, I was contacted out of the blue by officials from my hometown, Jiaoling. Calling it my "hometown" might be a Chinese thing, given that I have never lived there and did not visit the municipality until I was thirty years old. But that's where my father was from and where his ancestors had lived for the previous eight hundred or so years. So that's where my roots lie, even if I don't have a strong personal connection to the place.

The officials were planning to build a park along the Shiku River that they hoped would become a major tourist destination. Their plans called for the installation of new sculptures and statues, including one of me, since I was considered a relatively famous person among those with ties to the city. (Yes, the isolated hamlet I visited in 1979—without so much as a hotel or even a paved road—is now a bustling city that's about to have its very own waterfront park.) Although I was flattered by the offer, I suggested a somewhat different approach: Rather than doing a statue of me—a fairly ordinary-looking guy—perhaps they could focus, instead, upon a more extraordinary-looking shape, a Calabi-Yau manifold. That was the first round of our conversation, as both sides needed a bit of time to think things through. Our discussions have evolved considerably since then, and I have reason to believe this venture will soon go forward. (In China, once government leaders decide to do something, they can move much more quickly than their counterparts in the United States.)

At roughly the same time, I heard from Andy Hanson, a computer scientist at Indiana University, who had started his career several decades before as a physicist working in string theory and general relativity. Hanson had been a postdoc at IAS in 1971, after receiving his PhD from MIT, when I was also there, after graduating from Berkeley. At Princeton in those days, I spent an inordinate amount of time talking with Nigel Hitchin and others about the Calabi conjecture—a proposition that, as I've said, we considered "too good to be true." It's possible that some of those conversations rubbed off on Hanson, for he has since become the world's foremost creator of Calabi-Yau manifold visualizations, among his many other pursuits.

In 1999, Hanson produced illustrations of Calabi-Yau manifolds for Brian Greene's best-selling book *The Elegant Universe,* and four years later, he also contributed animations for the popular NOVA television program of the same name. One of his Calabi-Yau images graced the cover of a book I coauthored, *The Shape of Inner Space,* and I've relied upon his images in many lectures over the years. All this is a roundabout way of saying that Hanson was just the guy I needed to talk to.

Hanson told me, coincidentally, that he had teamed up with the Baltimore-based sculptor Bill Duffy in the hopes of mounting a four-foot-diameter sculpture of a Calabi-Yau space—a three-dimensional rendering of a six-dimensional manifold composed of either stainless steel or bronze—in a courtyard at his university. Their designs are quite far along, though Hanson has not yet secured final approval. Perhaps I'm biased, but I assume it's just a matter of time before the artistic and educational merits of this undertaking become apparent to the academic powers that be.

I told Hanson about the Jiaoling proposal—a bronze casting of a

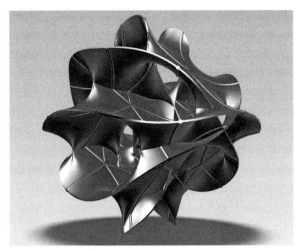

A model of the proposed Calabi-Yau sculpture for Indiana University created by Andrew J. Hanson of Indiana University and the sculptor William F. Duffy. (Design by Andrew J. Hanson, Indiana University, with William F. Duffy, Sculptor.)

Calabi-Yau manifold that would be roughly *sixteen* feet high, a veritable (as well as sizable) paean to geometry. His designs would, of course, be instrumental to that project. And the idea seems to be spreading: The Mathematical Sciences Center I helped start at Tsinghua University in Beijing appears to be interested in having a Calabi-Yau representation placed outdoors in the school's mathematics quadrangle. Hanson, in fact, returned in the spring of 2018 from a trip to China, where he met with my friend S. Y. Cheng, an associate director of the math center, to seek out the most propitious spot on campus to erect such a structure. Meanwhile, my colleagues and I are giving serious consideration to placing a sculpture at the Tsinghua-affiliated mathematics convention center I founded in the resort city of Sanya on Hainan Island.

So it's possible that before long, facsimiles of these manifolds will be sprouting up in various places. Of course, string theory postulates that at every possible spot you could point to in space—everywhere you go, everywhere you touch—there is a tiny, six-dimensional Calabi-Yau manifold, completely concealed from view, that still exerts a powerful influence on the physical world. This, according to string theory, is a vital—as well as ubiquitous—feature of the hidden universe. And if that premise is correct, these sculptures will represent one step toward making the invisible visible. And even if that bold assertion of string theory is not borne out, I still think that Calabi-Yau manifolds are beautiful. Beyond that, they have already proved their importance, countless times over, in both math and physics.

I was struck by the elegance of Calabi's equations during my first year of graduate school when I was a mere twenty years old—decades before I'd seen any of Hanson's alluring drawings or three-dimensional models. Mathematicians have always struggled to get nonpractitioners to share the passion and excitement they feel about their oftentimes esoteric field. I'm not naïve enough to think that these sculptures will finally communicate something that mathematicians have tried, unsuccessfully, to convey to the general public over the decades and, indeed, centuries. But these works, which fuse art and mathematics, could at least spark some interest. And perhaps a young boy or girl, out exploring a new park with his or her parents, might become curious about an unusually shaped object that's like nothing they've ever seen before—a by-product of nonlinear differential equations they've surely never seen before—in the same way they may be mesmerized by whirlpools that spontaneously appear in the nearby Shiku River (another example of a nonlinear phenomenon).

Some of these children might even be motivated to read the inscription at the base of the statue, which will describe the proof of the Calabi conjecture and the ways it has affected our thinking about the universe, elementary particle physics, and multiple areas of math, including differential and algebraic geometry and number theory. In other words, a lot has come from a little-known problem in geometry—far more than could have been predicted when Calabi posed his conjecture sixty-five years ago. If interest in mathematics, once kindled, grows from there, some of these young passersby might eventually gravitate toward the profession that I, too, was pulled toward as an inquisitive preteen. And even if just a single youth is drawn by this artwork into the study of math, that could be significant because in my field, one person—endowed with a bit of talent, drive, and luck—can make a difference.

Poincaré's Dream
On a long and silent night,
an equation resonated throughout the land.
Inspiration taking aim at an age-old conundrum,
revealing a path to the summit that for a century defied us.

Such beauty and splendor,
how did it come to be?
All due to graceful calculations,
transported upon a cosmic wave.

From a vantage high in the sky,
you found the elegance of topology,
and grasped the singularities of geometry,
leaving just one daunting pass to overcome.

Abstruse language, dreamlike deductions,
which drew upon the sum of your learning.
It took a single piercing estimation
to unfurl the complexity of the spheres.

With that, a light shone upon this earth
leaving nature's radiance unveiled.
Brightness for the rest of us to bask in,
and reassurance for the next faltering steps.

—*Shing-Tung Yau, 2006*

Ode to Space-Time
Why must time march on as it inexorably does?
And why do living things proliferate so?
Does not every drop of water share a common source,
just as mind and substance coexist in the same world?

Time passes by, swiftly, silently, never to return.
And the sky goes on, forever, with no limits in sight.
The universe continuously evolves, as do black holes,
space and time merging into one seamless whole.

So immense our puzzling universe appears,
yet how beautiful, inexpressible, the origin of truth is.
The great thinkers labor with patience and persistence,
meticulously fashioning a quantum view of space-time.

Filling in a picture that melds the large and small,
a view that conjoins the infinite to the infinitesimal.
Learning is but a peep into the vast unknown,
like dipping a cup into a boundless ocean.

—*Shing-Tung Yau, 2005*

INDEX

Note: An italicized page number indicates a figure.

Abhyankar, Shreeram, 80
Academia Sinica, 267–68
Academy of Mathematics and Systems Science, 170
acceleration, 53
Ahlfors, Lars, 129
Albert Einstein Chair, 84, 185, 267
algebraic geometry, 205
algebraic topology, 48, 136
Allen, Woody, 176
American Journal of Mathematics (journal), 87
American Mathematical Society (AMS), 165, 195
Ampère, André-Marie, 97
AMS. *See* American Mathematical Society
Annals of Mathematics, 60
Argonne National Laboratory, 177–79
Arkani-Hamed, Nima, 273
Armstrong, Neil, 45
Arnowitt, Richard, 160
Artin, Emil, 209
Artin, Michael, 209
Asian Journal of Mathematics (journal), 217, 243

Atiyah, Michael, 76, 113, 118, 125, 153, 155, 227, 254
Atkinson, Richard, 168
Aubin, Jean-Pierre, 143
Aubin, Thierry, 112

Bailey, David, 93
Bando, Shigetoshi, 159
Bartnik, Robert, 155, 161
Becker, Katrin, 203–4
Becker, Melanie, 203–4, 207
Bedford, Eric, 105
Beijing, 136, 148, 150, 220, 252
Bennett, Bruce, 91
Benney, David, 165
Benter, William, 216
Berkeley, 37, 50, 61, 70, 82–83, 133; acceptance to, 40–41; Center for Pure and Applied Mathematics, 148–49; education at, 64–65; math department at, 46–47; protests at, 54–55
Bers, Lipman, 130
Binglin, Gu, 253, 254
Bing topology, 153

biology, 209
Birkhoff, Garrett, 209
Birkhoff, George David, 209
Bloom, Harold, 185
Blum, Manuel, 65
Bôcher, Maxime, 190
Bochner, Salomon, 121
Bogomolov, Fedor, 111
Bogomolov-Miyaoka-Yau inequality, 111
Bombieri, Enrico, 143
Bondi mass, 160
Bonnet, Pierre, 50–51
Borel, Armand, 135, 143, 156, 164, 168–69
Bott, Raoul, 147, 188, 191
boundary value problems, 100
Bourguignon, Jean-Pierre, 86, 125, *128*, 143, 226
Bramson, Maury, 92–93
branes, 203–4
Brauer, Richard, 59
Brenner, Michael, 256
Brigham Young University, 93
British Consulate, 125
Brody, Elmer, 50
Browder, Bill, 154
Brown University, 256
Bryant, Robert, 143, 193

Calabi, Eugenio, 62–63, 64, 90, 95, 100, 114–15, 153, 164, 280
Calabi conjecture, 64, 68, 76–77, 80, *101*, 173, 278; counterexample to, 85–86, 95, 115–16; proof of, 109–10, 114–15, 134
Calabi-Yau manifolds, 175, 175–76, *176*, 191, 199, *199*, 246, 257; classification of, 206; Euler numbers and, 177–79; mirror manifolds of, 204; number of, 177; sculptures, 278, *279*; submanifolds of, 204, 205
Calabi-Yau theorem, 116
Caltech, 180, 196
Candelas, Philip, 173–74, 176, 178–79, 200–202, 203
Canton, 138, 139, 170–71
Cao, Huai-Dong, 159, 166, 182, 242–45
Casselman, Bill, 130

Cauchy, Augustin-Louis, 127
Cayley, Arthur, *201*
Cayley-Salmon theorem, *201*
Center for Mathematical Sciences and Applications (CMSA), 255–56
Center for Pure and Applied Mathematics, Berkeley, 148–49
Cha, Jimin, 214
Chan, Gerald, 221
Chan, Ronnie, 89, 216, 221, 223–24
Chan, Shu Kui, 23
Chandrasekharan, Komaravolu, 159
Chang, Kung-Ching, 222, 224–25, 267
Cheeger, Jeff, 143
Chen, Jiangong, 219
Chen, Jingrun, 150
Chen, Thomas, 216
Chen, Xiuxiong, 192–93
Chen, Zhili, 227–28
Cheng, S. Y., 83, 93, 97, 106, 122, 135, 152, 195, 216–18, 230, 247, 279; friendship with, 17, 66–67
Chern, S. S., 41, 52, 54, 60, 73, 83, 88–89, 103, 121, 153, 164, 213–14, 218, 246–47, 262–63; career of, 232; on China, 149–50; death of, 172, 231; Hua and, 91–92, 150, 171–72; on ICCM, 226; relationship with, 133–34, 149, 162–63, 172, 181, 195–96, 215, 222, 229, 231, 267–68; studying with, 65–66, 71–72; symposium organized by, 148–49
Chern Award, 195
Chern Institute of Mathematics, 262–63
Chern-Simons theory, 84
China, 2, 69–70, 136, 182, 207; Chern on, 149–50; citizenship, 194; culture in, 140; ICM in, 230; literature of, 21; mathematics in, 213–15, 266–67; Olympic Games in, 220; philosophy in, 274–75; politics in, 230; research culture in, 270–72; travels in, 150–51, 211–12; United States and, 80, 140, 261–62, 265, 273, 275–76; Yau, I., in, 211–13; Yau, M., in, 211–13, 217
China-U.S. Physics Examination and Application (CUSPEA), 165

Chinese Academy of Sciences, 136, 170, 181, 220, 267–68
Chinese Mathematical Society, 224, 225, 228, 266
Chinese Ministry of Education, 227
Chinese New Year, 10–11, 24
Chinese University of Hong Kong (CUHK), 34, 38–39, 74, 213
Chiu, Chen Ying, 2–16, 137, 155, 197; death of, 24–28, 198
Chiu, Isaac, 3, 163, 164, 185, 197; birth of, 158; in China, 211–13; relationship with, 167–68, 209; studies of, 209
Chiu, Michael, 3, 185, 191–92; birth of, 164–65; in China, 211–13, 217; relationship with, 209; studies of, 209
Chow, Ben, 159, 182
Chow, H. L., 35, 37
Chow, Ping Wah, 216
Chow, Wei-Liang, 94
Christmas, 56–57, 113
Christodolou, Demetrios, 185
Chu, Paul, 232
Chui, Siu-Tat, 34, 100
Chung, Fan, 256
Chung, Kai Lai, 91
Chung Chi College, 8, 24, 34, 36, 59
circles: lines and, 49; squaring, 21
Clay Mathematics Institute, 242, 243
Clemens, Herbert, 258
Clinton, Bill, 208
closed surfaces, 126–27
CMSA. *See* Center for Mathematical Sciences and Applications
Coates, John, 247, 273
Cohen, Paul, 157
Communications on Pure and Applied Mathematics (publication), 113, 182–83
"Compact Three-Dimensional Kähler Manifolds with Zero Ricci Curvature," 177
complex geometry, 67
complex manifolds, 91
Confucius, 3, 14, 26, 47, 67, 212, 249–50, 270, 274
Connelly, Robert, 126
Connelly sphere, *128*

Connes, Alain, 163
continuous symmetry, 85
Conway, John, 268
cooking, 46
Cornell, 94, 159
Courant Institute, 35, 87, 103, 105, 161, 182
Crick, Francis, 207
critical point theory, 78
cube, spheres and, 48, *49*
CUHK. *See* Chinese University of Hong Kong
Cultural Revolution, 141, 151, 165, 171, 220, 221
Cunningham, Merce, 185
curvature: general relativity and, 123; in geometry, 52–53; negative, 58; in physics, 53, 61; positive, 58; positive scalar, 126; of spheres, 58; zero, 58. *See also* Ricci curvature
CUSPEA. *See* China-U.S. Physics Examination and Application

Dazhong, Wang, 254
Dedekind, Richard, 35, 257
Dedekind cut, 35
Dehn, Max, 119
Dehn's lemma, *119*
de la Ossa, Xenia, 200
deLeeuw, Karel, 129
Department of Energy, 100, 190
Deser, Stanley, 123, 160
Diamond, Jared, 185
Diao Yu Tai movement, 69–72, 79
differential equations, 48, 70; nonlinear partial, 52, 86; ordinary, 36; partial, 50–51
differential geometry, 48, 90, 152
Ding, Shisun, 162–63, 183–84
Dirichlet, Peter Gustave Lejeune, 100
Dirichlet problem, 100, 101, *101*
discrete symmetry, 85
Dixon, Lance, 199
Donaldson, Simon, 154, 192, 264–65
Donoho, David, 186–87
donuts. *See* torus
Dopplereffekt, 176
The Double Helix (Watson), 207

Douglas, Jesse, 119
drug use, 55–56
duality, 199–200
Dubins, Lester, 65
Duffy, Bill, 278, 279

Edelman, Marian Wright, 185
education, 7, 13, 14; American-style, 36; at Berkeley, 64–65; physical, 19
Einstein, Albert, 50–51, 53, 62, 75, 115, 175
Elberg, Sandy, 46
elementary calculus, 84
Ellingsrud, Geir, 202
elliptic curves, 257
English language, 13–14, 17, 32
Enneper surfaces, *81*
entropy formula, 240–41
enumerative geometry, 200–201, 205–6
estimates and estimation, 86, *101*, 104
Euclidean geometry, 20, 32
Euclidean quantum gravity, 131
Euler, Leonhard, 52, 57, 63, 126–27, *178*, 256
Euler numbers: Calabi-Yau manifolds and, 177–79; creation of, *178*
European Mathematical Society, 226
exotic spheres, 84–85, 89

Fairchild Fellowship, 196
Fengmian, Lin, 3
Fermat's Last Theorem, 225
Fields Medal, 47, 74, 91, 114, 119, 143, 154, 160, 164, 186, 208, 246, 258; receiving, 166–67
financial crisis of 2008, 252
Finland, 129–30
first string revolution, 174
Fischer, Arthur, 61, 62
Fischer-Colbrie, Doris, 143
flexible polyhedrons, 127, *128*
Flexner, Abraham, 75
four-dimensional manifolds, 169–70
Frankel, Ted, 83
Frankel conjecture, 134
Franklin, Rosalind, 207
Freedman, Michael, 153, 169, 181, 185–86, 236

Fu, Jixiang, 206
Fudan University, 171
Fujian Province, 4

Gansu Province, 211
Gao, Zhiyong, 161–62
Garabedian, Paul, 186
Gårding, Lars, 148
Garland, Howard, 84
gauge theory, 201
Gauss, Carl Friedrich, 35, 50–51, 63, 129, 256, 274
Gauss-Bonnet theorem, 51–52
general relativity, theory of, 53, 173, 251; curvature and, 123
genius, 38
genus, 49
geometric analysis, 52, 130, 249; lectures on, 136; potential of, 118
geometric space, 98
geometrization conjecture, 259
geometry: algebraic, 205; complex, 67; curvature in, 52–53; differential, 48, 90, 152; enumerative, 200–201, 205–6; Euclidean, 20, 32; Kähler, 62; nature and, 50; projective, 71–72; topology and, 48, 49, 51–52, 59
Gepner, Doron, 199
Geroch, Robert, 89–90, 122, 160
Gibbons, Gary, 131, 251
Gieseker, David, 77, 111
Gilbarg, David, 89
Givental, Alexander, 202–3
Gleason, Andy, 92, 115, 191
Glimm, James, 186
Goldbach conjecture, 150
Gong, Sheng, 172
Gordon, Cameron, 120
Gore, Al, 208
graduate students, 106, 143, 155, 159, 169, 182, 185, 191
Graham, Ronald, 226
Grassmanians, 38
gravity, 62; Euclidean quantum, 131
Great Famine, 141
Green, Leon, 83
Green, Michael, 175, 177
Green, Paul, 200

green card, 65
Greene, Brian, 174, 190–91, 198–201, 205–6, 273, 278
Greene, Robert, 90, 107, 195
Griffiths, Phillip, 112, 114, 153, 162, 165, 168, 195; on politics, 247
Gromoll, Detlef, 60
Gromov, Mikhail, 98, 126, 164
Gross, David, 229–30, 273
Gross, Dick, 191
group theory, 68–69
Gruber, David, 245–46
Gu, Xianfeng, 176, 199, 205, 256
Guanghou, Zhang, 150
Guowei, Wang, 110

Hainan, 226
Hales, Thomas, 268
Hamilton, Richard, 145, 158, 169, 182, 276; Perelman and, 243–44; on Poincaré conjecture, 259; relationship with, 237–38; on Ricci flow, 237–38; at UCSD, 180, 188
Hamilton, William Rowan, 156
Hangzhou, 151, 218
Hanson, Andrew, 115, 278–80
Hardy, G. H., 39
Harish-Chandra, 147
Harvard, 72, 114, 135; during financial crisis, 252; hiring at, 188; history of, 189; mathematics at, 189–90; Strominger at, 206
Harvard, John, 189
Harvey, F. Reese, 204
Hatfield, Brian, 180
Hawking, Stephen, 124, 130, 132, 143, 229, 251
Hearst, William Randolph, III, 252
Helton, Bill, 185
Herman, Michael, 266
Hermitian-Yang-Mills equations, 158
Higgs boson, 269
Hilbert, David, 115, 129, 148, 256, 268
Hildebrandt, Stefan, 129, 143, 155
hippies, 55, 61
Hironaka, Heisuke, 91, 114, 147, 191
Hirzebruch, Friedrich, 67
Hitchin, Nigel, 76, 85, 95, 113, 125, 278

HKUST. *See* Hong Kong University of Science and Technology
Hodge, William, 118
homological mirror symmetry, 205
Hong Kong, 4–5, 31, 41, 70, 82, 83, 139, 182, 215
Hong Kong College, 23
Hong Kong Mathematical Society, 230
Hong Kong Rotary Club, 216
Hong Kong University of Science and Technology (HKUST), 215
Hopf, Heinz, 52
Hörmander, Lars, 148
Horowitz, Gary, 160, 162–63, 173, 174
Hsiang, Wen Chiao, 87–88, 155, 165–66
Hsiang, Wu-Chung, 74, 79–80, 85, 102, 154, 165, 268
Hsiang, Wu-Yi, 66, 73, 88, 134, 267–68
Hsinchu, 212, 217
Hsiung, Chuan-Chih, 195
Hu, Bei-Lok, 79, 124
Hu, Guoding, 218
Hua, Loo-Keng, 32, 39; Chern and, 91–92, 150, 171–72; illness of, 171; Wu, W., and, 136–37
Huangpu River, 151
Huisken, Gerhard, 169

IAS. *See* Institute for Advanced Study
ICCM. *See* International Congress of Chinese Mathematicians
ICM. *See* International Congress of Mathematicians
IHES. *See* Institut des Hautes Études Scientifques
IMS. *See* Institute of Mathematical Sciences
IMU. *See* International Mathematical Union
Institut des Hautes Études Scientifques (IHES), 125, 163
Institute for Advanced Study (IAS), 72, 143–44, 163, 184; history of, 75; permanent position at, 147; philosophy of, 75–76; research at, 155; study at, 80–81
Institute of Mathematical Sciences (IMS), 216, 217

integers, 35
Intel Science Talent Search, 209–10
International Congress of Chinese Mathematicians (ICCM), 225, 227; Chern on, 226; support of, 228
International Congress of Mathematicians (ICM), 101, 129, 147–48, 207, 228; in China, 230
International House (Berkeley), 46
International Mathematical Union (IMU), 159, 195, 220, 228, 247
irrational numbers, 35

Jacobian conjecture, 184
Jaffe, Arthur, 190
Japan, 70
Ji, Lizhen, 204–5
Jiaoling Country, 137, 139, 277
Jockey Club, 215
Johns Hopkins, 94, 147, 151, 152
Johnson, Samuel, 121
Jost, Jürgen, 155, 159
Journal of Differential Geometry (journal), 60, 153–55, 195, 236, 240
juvenile delinquency, 15–16

Kadison, Richard, 50
Kähler geometry, 62
Kähler manifolds, 156, 174, 206
Kao, Charles, 214, 215
Karlin, Sam, 92
Ka-shing, Li, 216, 221
Katz, Nick, 156
Katz, Sheldon, 200
Kazdan, Jerry, 112
Kepler, Johannes, 267–68
Kettering, Charles, 148
Kirby, Robion, 154
Kiremidjian, Garo, 91
Kisin, Mark, 252–53
Kleiner, Bruce, 242, 243
Klingenberg, Wilhelm, 129
Knight, James, 37
Kobayashi, Shoshichi, 41, 67
Kodaira, Kunihiko, 73, 118
Kohn, Joe, 155–56, 172
Kontsevich, Maxim, 202
Kuiper, Nicolaas, 125

Kuok, Robert, 216
Kuratowski, Kazimierz, 92

Lam, Ping-Fun, 78–79, 83
Lam, Tsit Yuen, 46, 66, 71–72
Langlands, Robert, 129
Langrange, Joseph-Louis, 81
Large Hadron Collider, 270
Lawson, Blaine, 48, 69, 85, 89, 143, 153, 161–62, 164, 204; work with, 59–61
Lee, K. Y., 30
Lee, Pak Win, 33
Lee, Tsung-Dao, 41, 80, 165, 224, 269, 271
Lee Foundation, 216
Lerche, Wolfgang, 199
Leung, Conan, 205
Leung, Yeuk Lam, 2–11, 23–24, 27–28, 42, 151–52; death of, 197; funeral of, 197–98; illness of, 196; relationship with, 29–30, 198
Li, Jun, 171, 191, 206, 226, 238
Li, Peter, 122, 135, 143, 181, 237–38
Lian, Bong, 202, 258
Liebniz, Gottfried, 32
Lin, Chang-Shou, 166, 225
linear algebra, 35
linear equations, 53, 86–87
lines, circles and, 49
Ling, Dao-Yang, 24
Little Italy, 106
Liu, Ai-Ko, 213
Liu, Chao-shiuan, 212
Liu, Jun, 256
Liu, Kefeng, 202, 254, 262
Liu, Melissa, 251
Li-Yau inequality, 181–82, 238
Lo, Yang, 150, 170, 180–81, 217–19, 225, 231
Lobastova, Nadejda, 241–42
London Mathematical Society, 125
Long, Donlin, 147
Lott, John, 242–43
Lu, Qi-Keng, 137, 144, 194
Lu, Yongxiang, 220–21, 223
Lurie, Jacob, 252–53

Ma, Zhi-Ming, 229
MacArthur Fellowship, 184, 185

Mackenzie, Dana, 259
Mackey, George, 191
Mac Lane, Saunders, 133, 150
Mak, Frances, 94
malnourishment, 29
"Manifold Destiny" (Nasar and Gruber), 245–46
Manuscripta Mathematica, 126
Mao Zedong, 79, 124, 212, 253
martial law, 163
Martin, Francesco, 176
Massachusetts Institute of Technology (MIT), 50, 72, 113, 165; Tian at, 192–93; Yau, Y. Y., at, 188
Mathematical Aspects of String Theory, 180
Mathematical Sciences Research Institute (MSRI), 133
mathematics, 19–20; in China, 213–15, 266–67; fields in, 51; at Harvard, 189–90; importance of, 256–57; love of, 35; nature and, 50; physics and, 63
Mathematics Research Center (MRC), 186
Mather, John, 266
Max Planck Institute, 155
Maxwell, 50
Mazur, Barry, 191
Meeks, Bill, 119, 122, 259; working with, 118–20
Melrose, Richard, 135
Mencius, 3
Mengzi, 3
Mid-Autumn Moon Festival, 11
Miles, John, 187
Millennium Prizes, 243
Mills, Robert, 113, 269
Milnor, John, 57, 69, 78, 119, 126, 147, 153, 238
minimal surfaces, 80–81, 81, 118; positive mass conjecture and, 123; topology and, 119–20
Minkowski problem, 97–98, 105
mirror conjecture, 202; proof of, 203
mirror manifolds, 199; of Calabi-Yau manifolds, 204
mirror symmetry, 199–202, 204, 206, 257; homological, 205

Misner, Charles, 160
MIT. *See* Massachusetts Institute of Technology
Möbius strip, 49
Mogao Caves, 211
Moh, Tzuong-Tsieng, 79, 80, 184
Mok, Ngaiming, 156, 157, 172
Mong, William, 216
Monge, Gaspard, 96–97
Monge-Ampère equations, 97, 163, 256; solutions to, 104
Moore, Calvin, 133
Morel, Sophie, 252–53
Morgan, John, 242
Mori, Shigefumi, 134–35
Morison, Samuel Eliot, 189
Morningside Center of Mathematics, 221–22
Morningside Lifetime Achievement Medal, 226
Morningside Medal of Mathematics, 222, 225
Morrey, Charles, 48, 50–51, 54, 56, 65, 70, 86, 98
Morse, Marston, 78, 153
Morse theory, 78
Morse Theory (Milnor), 78
Moser, Jürgen, 103, 105, 159–60, 166
MRC. *See* Mathematics Research Center
MSRI. *See* Mathematical Sciences Research Institute
Mumford, David, 111–12, 114, 147, 156, 182, 191, 256

Nankai University, 213, 215, 218, 229, 262
NAS. *See* National Academy of Sciences
Nasar, Sylvia, 245–47, 250
Nash, John, 245
National Academy of Sciences (NAS), 150; establishment of, 265; nomination at, 195
National Medal of Science, 91, 208
National Science Foundation, 52, 65, 133, 193–94
nature, 50
negative curvature, 58
New Asia College, 37

Newton, Isaac, 32, 50–51
Newton's method, *101*
New York City, 80, 83, 103, 105–6, 144
New Yorker (magazine), 257–58
Nirenberg, Louis, 87, 90, 101–2, 113, 143, 153, 166
Nixon, Richard, 80, 194
Nobel Prize, 17, 41, 84, 177, 188, 214, 224, 269, 271
nonlinear analysis, 54
nonlinear equations, 53–54
nonlinear form, 50–51
nonlinear partial differential equations, 52, 86
Northrop Grumman, 106
number theory, string theory and, 258

Olshen, Richard, 186–87
operator algebra, 50
ordinary differential equations, 36
orientable surfaces, *49*
Osgood, William Fogg, 190
Osserman, Robert, 87, 89, 94, 102, 168
Overbye, Dennis, 250–51

Palis, Jacob, 229
Palo Alto, 91
Papakyriakopoulos, Christos, 119
Paris, 127–28
Parkes, Linda, 200
Parkinson's disease, 66
partial differential equations: nonlinear, 52, 86; physics and, 50–51; solutions to, 50–51
particle physics, 269
Peirce, Benjamin, 109, 190
Peking University, 162–63, 222–23
Penrose, Roger, 129, 132, 143, 251
Perelman, Grisha, 240–42, 246, 258, 259; Hamilton, R., and, 243–44
perfect numbers, 190
Perry, Malcolm, 143, 160, 161
Phillips, Ralph, 94
Phillips, Tony, 83
Physical Dynamics, 132
physical education, 19
physics: curvature in, 53, 61; mathematics and, 63; partial differential equations and, 50–51; particle, 269; topology and, 61
Pig House, 28, 30, 32–33
Plateau, Joseph, 81, *81*
Plesser, Ronen, 199–201, 205–6
poetry, 11–12, 27, 110, 280–82
Pogorelov, Aleksei, 105
Poincaré, Henri, 52, 63, 233–34, 236, 256
Poincaré conjecture, 107, 158, 240; Hamilton, R., on, 259; proof of, 144, 186, 232, 238, 242, 244, 259
Poincaré proof, 182
Poland, 163
Polchinski, Joe, 203
polyhedrons: classification of, *178*; flexible, 127, *128*
Portsmouth Square, 70
positive curvature, 58
positive mass conjecture, 122, 125–26, 131, 160–61, 251; minimal surfaces and, 123
positive scalar curvature, manifolds of, 126
poverty, 6, 33
Preissman, Alexandre, 57–58
Preissman's theorem, 58–59, 64
Princeton, 72, 76, 102, 106, 121, 152, 156–57, 267
Princeton Plasma Physics Laboratory, 99–100
Princeton University Press, 152
Proceedings of the National Academy of Sciences, 113
projective geometry, 71–72
proof by contradiction, 77
Protter, Murray, 148
public exams, 14–15
Pui Ching Middle School, 16–17, 23, 30
Purdue, 80
Pure and Applied Mathematics Delegation, 150
pyramid, *49*
Pythagoras, 50

Qing Dynasty, 117, 274
quantum mechanics, 173
quasi-local mass, 252
quintic, 39, 200, 203

rational curves, 257
rational numbers, 35
real numbers, 35
Red Guards, 212
Ribet, Ken, 127–28
Ricci curvature, 161–62, 259; tensor, 62, 64
Ricci flow, 145, 158–59, 169, 181; Hamilton, R., on, 237–38; three-dimensional manifolds and, 239
Rice University, 121
Rieffel, Marc, 46
Riemann, Bernhard, 35, 62, 129, 256, 274
Riemann hypothesis, 68
Riemann zeta function, 77
Roan, Shi-shyr, 182
Romance of the Three Kingdoms, 168
Romeo, Alfa, 122
Rosenblatt, Murray, 187
Rosenthal, Haskell, 65
Rosovsky, Henry, 168
Roy, Andrew Tod, 146–47
Roy, J. Stapleton, 146–47

Sabitov, Idzhad, 127
Saint-Donat, Bernard, 128–29, 252
Salaff, Stephen, 36–37, 39–40; friendship with, 55–56
Salmon, George, 201
Samuelson, Hans, 168
San Diego, 168–69, 185. *See also* University of California, San Diego
San Francisco, 44, 102
Sarason, Donald, 40, 45
Sarnak, Peter, 157
Schmid, Wilfried, 191, 196, 247
Schoen, Rick, 89, 122, 124–26, 135, 138, 158, 164, 169, 182, 251, 259; collaborations with, 180; at UCSD, 180
Schubert, Hermann, 200–201
Schultz, Reinhard, 84–85
Schur, Issai, 59
Schwarz, John, 175, 177, 196
Science Talent Search, 272
Segal, Irving, 50
Seidenberg, Abraham, 65
Selberg, Atle, 147
Sepulveda, 109

Serre, Jean-Pierre, 156
Severi conjecture, 112, 116
Shantou, 2, 4, 182
The Shape of Inner Space (Yau, S. T., and Nadis), 278
Shatin, 8, 18, 28, 33
Sheng, Ping, 79, 80
Shi, Wan-Xiong, 171, 191
Shintani, Takuro, 77
Shintani zeta function, 77
Shioda, Tetsuji, 129
Shisun, Ding, 264
Shore, Dinah, 88
Siegel, Carl Ludvig, 156
Simon, Leon, 133, 185, 259
Simons, Jim, 72, 82–84, 87–88
Simons Foundation, 252
simply connected, 234, 235
Singer, Isadore, 112, 113, 118, 122, 190; on Tian, 192–93
singularities, 119, 132, 237
Siu, Yum-Tong, 17, 130, 134–35, 143, 156, 165; Tian and, 183
Sloane, Neil, 268
Sloan Fellow, 112, 122
Sloan Foundation, 102
Smale, Stephen, 122, 236
Smarr, Larry, 123–24
Smith, Paul, 120
Smith conjecture, 119–20
Solidarity Movement, 163
Song, Jian, 223
Song Dynasty, 110
Spaces of Constant Curvature (Wolf), 60
Spanier, Edwin, 48, 54–55
special Lagrangian cycles, 204
spectrum of geometric space, 98
Spence, Michael, 188
spheres, 234; Connelly, 128; cubes and, 48, 49; curvature of, 58; defining, 53; exotic, 84–85, 89; 2-sphere, 234
squaring a circle, 21
Stallings, John, 236
Standard Model, 178–79, 269
Stanford, 87–88, 93, 103–4, 130; leaving, 168–69
Stark, Harold, 183
Steele, Michael, 145–46

Stein, Eli, 195
Stein, Gertrude, 233
Sternberg, Shlomo, 191
Stony Brook University, 72, 82, 84–85, 87, 126, 161–62
Streleski, Theodor, 129
string theory, 174–75, 176, 178–79, 205–6, 230, 249; contributions of, 179; limitations of, 179; number theory and, 258
Strominger, Andrew, 173–74, 177, 203–4, 229–30, 258; at Harvard, 206
Strominger, Jack, 209
Strominger-Yau-Zaslow (SYZ) conjecture, 204, 205, 249
Strømme, Stein Arild, 202
Sullivan, Dennis, 83, 185–86
Sun, Song, 192–93
superstring theory, 174
supersymmetry, 62–63, 174
Supersymmetry and Morse Theory (Witten), 154
symmetry, 62–63; continuous, 85
synergistic phenomena, 142
SYZ conjecture. *See* Strominger-Yau-Zaslow conjecture

tai chi, 37
Taiwan, 23, 70, 72, 212, 217
Tang Dynasty, 171–72
Tao, Terry, 273
Taoism, 10
Tate, John, 191
Taubes, Clifford, 191, 256
Taylor, Paul, 185
teaching, at Stony Brook, 84–85
Telegraph Avenue, 46
Terng, Chuu-Lian, 193–94
tetrahedron, 49, 178, *178*
Thanksgiving, 56
Thomas, Emery, 65
't Hooft, Gerard, 177
Thorne, Kip, 196
three-dimensional manifolds, Ricci flow and, 239
threefolds, 200
Thurston, Bill, 47, 99, 120, 130, 162–65, 241

Thurston's geometrization conjecture, 241
Tian, Gang, 171, 180, 191, 226, 242; at MIT, 192–93; relationship with, 192, 247–48, 264–65; Singer on, 192–93; Siu and, 183; work of, 193, 264–65, 267
Ticho, Harold, 187
time-symmetric case, 123
Todorov, Andrey, 114
Topological Methods in Algebraic Geometry (Hirzebruch), 67, 129
topology, 88, 177–78; algebraic, 48, 136; Bing, 153; curvature in, 52–53; geometry and, 48, 49, 51–52, 59; minimal surfaces and, 119–20; physics and, 61
torus, 49, 78, 205, 234, 235; upright, 78
trapped surfaces, 132
Treibergs, Andrejs, 143
Trèves, François, 161
triangles, problems with, 20–21
trivial case, 58–59
Tsinghua University, 39, 213, 217, 227, 252, 253, 263, 272, 279
Tsui, Daniel, 17
Turner-Smith, Ronald Francis, 59
tutoring, 31–34
twin prime conjecture, 184
two-dimensional surfaces, 236
2-sphere, 234

UCSD. *See* University of California, San Diego
Uhlenbeck, Karen, 113, 143, 158, 182, 185
United Kingdom, 31, 125
United States: China and, 80, 140, 261–62, 265, 273, 275–76; citizenship, 194; departure to, 42–43; visa for, 42, 82
University of California, San Diego (UCSD), 82–83; Hamilton, R., at, 180, 188; leaving, 187–88; politics at, 185–86; Schoen at, 180; staff at, 185; working at, 168–69
University of Chicago, 89, 150
University of London, 59

University of Maryland, 124
upright torus, 78

"Vacuum Configurations for Superstrings" (Candelas, Horowitz, Strominger, and Witten), 174
Vafa, Cumrun, 203, 257
van de Ven, Antonius, 111
velocity, 53
Vietnam War, 54–55, 70, 82
visa, for United States, 42, 82
von Dyck, Walther, 50–51

Walz, Anke, 127
Wang, Chin-Lung, 213
Wang, Hsien Chung, 94, 137, 139
Wang, Mu-Tao, 213, 252
Wang, Silei, 218
Wang, Yifang, 270
Waring problem, 150
Warner, Nicholas, 199
Watson, James, 207
Weil, André, 68, 129
Weinberg, Robert, 207
Weinstein, Alan, 65
Weyl, Hermann, 118, 269
Wheeler, John, 62, 124
white card, 125
Wiles, Andrew, 225
Witten, Edward, 154, 173–74, 177, 203, 229–30
Wolf, Joe, 60, 69
Wolff, Tom, 196
Wolf Prize, 249
Wong, Bun, 151, 152
Wong, Eugene, 69
Wong, Pit-Mann, 182–83
Wong, Y. C., 38
Woolf, Harry, 147
Wu, Hung-Hsi, 66, 72, 88, 90
Wu, Wenjun, Hua and, 136–37

Xiaopeng, Deng, 1, 219, 224
Xin, Zhou-Ping, 217
Xinjiang Province, 211
Xueqin, Cao, 117

Yamabe problem, 180
Yamaguchi, Satoshi, 257
Yang, Chen Ning, 41, 71, 80, 84, 113, 161–62, 213–14, 224, 227, 255, 267–69, 271
Yang, Paul, 82
Yang-Mills theory, 144, 154–55, 191, 269
Yat-sen, Sun, 1
Yau, Horng-Tzer, 256
Yau, Shing-Ho, 9
Yau, Shing-Hu, 2, 23, 196
Yau, Shing-Shan, 2, 31
Yau, Shing-Tung. *See specific topics*
Yau, Shing-Yue, 28, 31, 196
Yau, Shing-Yuk, 2, 17, 34, 42; death of, 167, 198; illness of, 146–47, 167
Yau, Stephen, 43, 163
Yau, Yu-Yun, 82, 83, 99, 102–4, 135, 150, 164, 207, 236–37; honeymoon with, 109; marriage to, 106–7, 109; meeting, 73; at MIT, 188; as mother, 210; pregnancy of, 151, 157–58; relationship with, 74, 93
Yau conjecture, 192
Yau Mathematical Sciences Center, 227, 255
Yellowstone, 107
Yin, Xiaotian, 176, 199, 205
YMCA, 45–46
Yosemite National Park, 100
Yuan, Qu, 47
Yuen Long, 5, 12
Yuxiu, Gu, 254

Zaslow, Eric, 204, 257–58
Zemin, Jiang, 218, 219, 221, 223, 229
Zeng, Ying-Cai, 31
zero curvature, 58
Zhang, Shou-Wu, 225
Zhang, Yitang, 183
Zheng, Fangyang, 171, 191
Zhong, Jia-Qing, 157, 183
Zhu, BangFen, 269
Zhu, Xi Peng, 242–45
Zurich, 159